CAMBRIDGE COUNTY GEOGRAPHIES

General Editor: F. H. H. GUILLEMARD, M.A., M.D.

T0175986

CAMBRIDGESHIRE

CAMBRIDGESHIRE

by

T. M^CKENNY HUGHES, M.A., F.R.S.

WOODWARDIAN PROFESSOR OF GEOLOGY

AND

MARY CAROLINE HUGHES

With Maps, Diagrams and Illustrations

Cambridge :
at the University Press
1909

CAMBRIDGE UNIVERSITY PRESS
Cambridge, New York, Melbourne, Madrid, Cape Town,
Singapore, São Paulo, Delhi, Mexico City

Cambridge University Press
The Edinburgh Building, Cambridge CB2 8RU, UK

Published in the United States of America by Cambridge University Press, New York

www.cambridge.org
Information on this title: www.cambridge.org/9781107651579

First published 1909
First paperback edition 2013

A catalogue record for this publication is available from the British Library

ISBN 978-1-107-65157-9 Paperback

NOTE

It is convenient to acknowledge here the assistance received from books and personally from some of their authors, instead of giving references in footnotes, or in the text.

Many of the old works on Cambridge and its surrounding district are very interesting reading, but recent writers, such as Mr J. W. Clark for the Town and University, and the Reverend Edward Conybeare for the County, have epitomised them, and to both these authors we are much indebted. Mr Conybeare's *History of Cambridgeshire* and *Rides round Cambridge* have been invaluable, and we are grateful to him for much kind help.

We have made use of Professor Skeat's books on Place-names and have also received valuable help from him in regard to local dialect words.

To the Misses Bull of Cottenham we are indebted for a list of ancient words and phrases, still used in their district, which, had space allowed, we should have liked to reproduce in full.

We also wish to thank Mr William Farren for advice on special points relating to birds and insects, and Miss Brinkworth and Mr Eric Titterington for lists of flowers.

Acknowledgements referring to the pictures will be found at the end of the list of illustrations.

We wish also to thank the Staff of the University Press for much kind attention and indulgence in our attempt to condense a large subject into a small space.

T. M^cKENNY HUGHES.

M. C. HUGHES.

CAMBRIDGE,
November 1909.

CONTENTS

		PAGE
1.	County and Shire	1
2.	General Characteristics. Position and Natural Conditions	7
3.	Size. Shape. Boundaries	10
4.	Surface and General Features	14
5.	The Fens	24
6.	Watershed and Rivers	35
7 a.	Geology and Soil	49
7 b.	Geology and Soil	61
8.	Natural History	69
9.	Climate	78
10.	People—Race, Dialect, Settlements, Population .	83
11.	Agriculture	89
12.	Forestry	94
13.	Special Cultivations	96
14.	Industries and Manufactures	101

		PAGE
15.	Mines and Minerals	109
16.	Fishing	114
17.	Shipping and Trade	117
18.	History	123
19.	Antiquities	132
20.	Architecture—(a) Ecclesiastical	153
21.	Architecture—(b) Military	172
22.	Architecture—(c) Domestic	177
23.	Communications: Past and Present	197
24.	Administration and Divisions of the County	204
25.	Roll of Honour of the County	209
26.	The Chief Towns and Villages of Cambridgeshire	222
Index		263

ILLUSTRATIONS

	PAGE
A page of Domesday Book	4
Thorney Road, Flooded, Whittlesea	8
The Nene between Peterborough and Guyhirne	13
Source of the Cam at Ashwell	17
Aldreth Causeway	20
Sutton Church	21
Peeling Willows	23
Meeting of the Burwell and Reach Lodes, near Upware	25
Section seen in Whittlesea Brickpit	27
Ancient Bone Skates	29
Burwell Lode	31
Peat digging, Burwell Fen	32
Peat at Burwell Fen	33
Old Bridge near Haslingfield	38
The Cam from King's College Bridge	39
The Cam at St John's College	40
The Lent Races. A Bump	42
The Cam at Upware	43
The Old Bedford River	45
The Nene at Foul Anchor	46
Wisbech at High Tide	47
Wisbech at Low Tide	48
Diagram Section from Snowdon to Harwich	50
Vertical Section	51
Diagram Section across Cambridgeshire	52
Roslyn Pit, Ely	55

PAGE

Ammonites biplex 58
The Chalk Quarry, Norman Cement Works . . . 60
Ice-scratched Boulder of Chalk, Roslyn Pit . . . 63
Head of Rhinoceros and of Hippopotamus . . . 65
Extinct Shells from Cambridge Gravel 67
The Swallow-tail Butterfly with its Caterpillar, Chrysalis
    and Food 73
Chopping Fen Litter, Reach 76
In the Fens. Coming back from Work . . . 85
Old Style Skating Champions. Turkey Smart and William
    See 87
The Woad Mill, Parson's Drove, Wisbech . . . 97
Woad Cultivation near Wisbech 99
Gathering the Strawberry Crop, Histon 102
Parchment Making, Sawston 104
The University Press 107
Part of Machine Room, University Press . . . 108
Pit in Chalk Marl, Bottisham Cement Works . . 111
The Port, Wisbech 118
Hundred Foot River, looking N.E., Mepal . . . 121
Coin of Allectus, found in the Cam Valley . . . 124
Archbishop Wilfrid installing St Etheldreda as Abbess of
    Ely 125
The Old West River, where the Causeway crosses it . 129
Prehistoric Implements, found in Cambridgeshire . . 135
Urus (*Bos primigenius*) found in the peat of Burwell Fen 137
Fleam Dyke, looking West 141
Urns. 1. British. 2. Late Celtic. 3. Roman 4. Saxon 146
Brooch, from Allington Hill 151
Fibula found in the Saxon cemetery at Haslingfield . 152
St Benedict's Church, Cambridge 155
Arch, St Benedict's Church, Cambridge 157
Prior's Door, Ely Cathedral 159

PAGE

Ely Cathedral from the Ouse 160
St Sepulchre's Church at the beginning of the Nineteenth
    Century 162
St Sepulchre's Church: Interior 163
Ely Cathedral, Nave East 164
Prior Crauden's Chapel, Ely 166
King's College Chapel, from the West 168
King's College Chapel, from the South . . . 169
The Great Court, Trinity College 171
Cambridge Castle and Castle Hill from the Huntingdon
    Road 173
St Peter's Church, Cambridge 174
Sawston Hall 175
Castle Ruins and Moat, Burwell 176
St John's College Dining Hall 177
An Old Farm on the Cam near Haslingfield . . . 178
Queens' College, Cloister Court 179
Madingley Hall 181
Priest's Hole, Sawston Hall 182
Senate House and University Library . . . 185
Peterhouse 186
Queens' College from the River 188
Christ's College, 1st Court 189
The Combination Room, St John's College . . . 190
Trinity College Library 191
Emmanuel College 192
Gate of Honour, Gonville and Caius College . . 193
The Fitzwilliam Museum 194
The Senate House 195
The Leys School 196
Worsted Street (Roman Road) 200
Jeremy Taylor 211
Thomas Hobson 212

                                                          PAGE
Statue of Sir Isaac Newton, Trinity College Chapel   .   218
King's College Chapel   221
Little Abington   222
St Sepulchre's Church, Cambridge   225
St John's College, Entrance Gate   227
Stourbridge Chapel   228
Ely Cathedral, Tower from South Side   233
Girton College   236
Newnham College   237
Newmarket, from an old print   243
Snailwell Church   247
Windmill, Swaffham Prior   248
Remains of two Churches at Swaffham Prior   .   .   249
Skating for the Championship at Swavesey   .   .   .   251
S.W. View of Thorney Abbey Church   252
Thorney Abbey   253
Denny Abbey near Waterbeach   254
The Market Place, Whittlesea   255
St Mary's Church, Whittlesea   256
Diagrams   259

MAPS

Basin of Great Ouse as compared with Cambridgeshire .   36
England and Wales, showing annual rainfall .   .   .   81
Map of Cambridgeshire showing the Dykes and places
        where antiquities have been found   133

The illustrations on pp. 17, 124, 247, are from photographs taken by the Rev. E. Conybeare; those on pp. 8 and 27, are from photographs of Mr John Crowson; those on pp. 13, 20, 21, 31, 38, 43, 45, 46, 48, 60, 76, 85, 104, 111, 121, 129, 141, 178,

200, 248, are by Mrs McKenny Hughes, taken especially for this book; those on pp. 25, 29, 32, 33, 42, 87, 251, are from photographs taken by Messrs Scott and Wilkinson; those on pp. 39, 40, 47, 155, 157, 159, 160, 163, 164, 168, 169, 171, 174, 177, 179, 181, 185, 186, 189, 191, 192, 193, 194, 195, 196, 218, 221, 225, 227, 233, 236, 237, 253, 255, 256, 259, are from photographs by Messrs Frith & Co.; the view on p. 55 is reproduced from a sketch by Canon Weston; those on pp. 58, 63, 65, 67, 135, 146, are from photographs taken by Mr W. H. Hayles and that on p. 73 is also by him from an original painting by Mr William Farren; the diagrams and sections on pp. 50, 51, 52, and the map, p. 133, are by Professor Hughes. The views on pp. 97, 99, 118, are from photographs by Messrs Lawrence Brown & Co.; that on p. 102, is by Messrs Mason and Basevi; those on pp. 107, 108, are supplied by the University Press; that on p. 137, by kind permission of the Society of Antiquaries; the figures on pp. 151, 152, are from drawings by Mrs Hugh Strickland; the view on p. 175, is reproduced by kind permission of Mr Huddleston; that on p. 190, is from a photograph taken by Mr E. Clennett; the illustration, p. 182, of "The Priest's Hole, Sawston Hall" is taken from Mr Allan Fea's book *Secret Chambers and Hiding Places*, by kind permission of the Author and Messrs Methuen & Co.; the view on p. 23, is taken from a photograph by Mr T. Bolton; the rest of the views are from old prints, or from pictures; the diagrams were compiled by Mr H. A. Parsons.

1. County and Shire.

We can learn much of the history of our country from the names of its counties. Many of them carry us back to the time before the different tribes and kingdoms had been welded together into one united England. So we find Essex, Middlesex, and Sussex still representing the ancient divisions of the East, Middle, and South Saxons, and Kent the kingdom of the Cantii. The early English people had a very complete system by which they planned out their land : each village had its share, sufficient for the needs of the community. From the tenth century it was found convenient to count a number of these village communities, connected by geographical or other reasons, as one "shire" for taxation and for raising fighting men for the King, under a "shire reeve" or sheriff, who was responsible for the levying of both men and money. "Shire" comes from the Old English word meaning to shear, because these divisions were shorn or cut off from old tribal districts. They were formed gradually, sometimes taking in a strip of country on the borders of two kingdoms; their boundaries changed from time to time according to circumstances, or as one tribe became powerful and enlarged its borders at its neighbour's expense.

After the Norman Conquest the Ealdorman or Earl, who was the chief man of the shire, was called Comes or Count. He was looked upon as a companion to the King and the district over which he presided was called a county. The Sheriff became Vice-Comes—deputy of the Comes.

Most of our counties which end in "shire" were only part of one of the early tribal divisions, and this is the case with Cambridgeshire. It was formed in the troublous times when the Danes were harrying this country. All the most important part of it was in East Anglia, but the western district was in Mercia, the Cam and Ouse forming a natural boundary between the kingdoms, and in the seventh, eighth, and ninth centuries, Cambridgeshire was the scene of fierce fights between the Mercians and East Anglians. We must remember this fact when we come to look into Cambridgeshire place-names and dialect, and seek from them an explanation of several curious facts about our county. The independence of the Isle of Ely dates back to the time when all that part of the county was the tribal settlement of the fenmen, or Girvii, who were allies of the East Anglians. The foundress of the Great Abbey of Ely inherited this land as her dower from their king, her first husband. She, being also an East Anglian princess, joined the interests of the two kingdoms and extended the influence of East Anglia in the Isle.

Cambridgeshire was never the home of one united tribe, nor as in Westmorland, did marked physical features suggest a name, but when a distinguishing title had to be given to this political unit, it was called after an important town within its borders first mentioned in the Old English Chronicle, A.D. 875 as Grante-brycge, while in 1010 the county makes its appearance in the same Chronicle as Grantabrycg-seir.

When, however, we come to look into the question of where this ancient town stood and how it came, in later times, to be called Cambridge, we meet with all sorts of curious and unexpected difficulties.

The town by the bridge over the Cam seems simple and obvious, but we find that this derivation will not do at all. We have in this enquiry an interesting example of the way in which guess-work at the origin of place-names is corrected by patient research into the evidence of ancient documents and the laws of word-making.

Some writers have suggested that Cam took its name from the Welsh adjective *cam*, crooked, or that it came from the substantive meaning, the bend, with rhyd = a ford—making " the ford at the bend." In the Itinera of the Roman historian Antoninus there is a town called Camboritum or Camboricum which must have been somewhere in this part of the country; so—the argument continued—the Romans adopted the Celtic name of the river or of the strong bend in the loop of which the modern town stands; they turned "rhyd" into "ritum," and Camboritum in time became Cambridge.

But these ingenious suggestions fall to the ground when we find that Cam was not used as part of the town

name until after 1400, and that the river was known in
the earliest times as the Granta. In 1130 the Church of
St Giles near the Castle is said to stand "Super Grantam

A page of Domesday Book giving a list of some of the lands
in Cambridgeshire belonging to Picot the Sheriff, A.D. 1081–86

fluvium." In Domesday Book we find mention of
Grentebrige. Professor Skeat tells us that in 1142 the

initial C first appears because the Normans mispronounced the "Gr," and avoided the use of these two letters and " br " in the same word. Gradually we get from Cantebrug to Canbrigge, Caumbrege to Cambryge, and sixteenth century scholars writing in Latin, deriving the name from the town, called the river Camus.

In the middle ages the Cam was le Ee, Old English for a river, and le Ree, and after 1372 the Canta. There is another Cambridge—a village in Gloucestershire, also on a stream now named Cam, which is called by a tenth century chronicler Cantbricge: the same writer calls the East Anglian Cambridge Grantanbricge. Mr Arthur Gray, who holds that the northern town and bank of the river were held by Mercia, says: " Cant or Cante seems to have been one of the many names applied to the river at Cambridge. Perhaps it is the same word as Kent or Kennett, a common river name. Cantebrige was apparently the name originally given to the town on the northern bank near the Castle and afterwards extended to the whole town."

Amidst this tangle of names one thing is certain, Cambridge is not derived from Camboritum and we may put the Celtic word Cam also out of our minds in this connection. The Welsh call Cambridge Caergraunt, but for how long they have done so is not clear, and we feel inclined to say with Camden " let others hunt after the derivation of this name." The second syllable in Cambridge does not necessarily mean a bridge, though we know of the Great Bridge by which the river was crossed at the Castle End. When this bridge was rebuilt

traces of a "firm pavement of pebbles" were found,
showing signs of an ancient ford there. The Scandinavian *bryggr* means a quay, and is the same thing as
hithe, so common in later times. In the early Sagas
bruggr is commonly used for landing places by the sea.
We have an example in our own country in Filey Brigg.
Invaders from the Continent landed on our shores and
crept up our rivers long before Roman times. It was
to keep off such attacks that, in the fifth century A.D.,
the Count of the Saxon Shore was appointed. So it
is along our waterways that we may look for the
earliest traces of such invaders. They landed on the
beach near the mouth of the Ouse and called it
Ouse-bech (Wisbech); they travelled up the river to
where gravel terraces offered a dry landing place, calling
it Waterbech or Landbech, according to the point of
view of settlers on the shore or travellers by the water.
At the great bend of the river, where it first cuts into
stiff clay or chalk, which geologists would call solid rock,
as opposed to surface deposits of gravel or mud, there, on
the river side of the Castle Hill we find a spot which
must have been important at any rate as far back as
Roman times. Here was a quay or bridge or both, and
here was an important dwelling and meeting-place for
the early inhabitants and settlers in the country. Here,
then, determined by natural conditions, our chief town,
from which the county takes its name, gradually grew
and gave a home to the great University which has made
it famous.

2. General Characteristics. Position and Natural Conditions.

When we come to consider the position of the county and the conditions affecting the occupation of the district by man, though Cambridgeshire has no sea shore nor any ranges of mountains to suggest necessary or convenient boundaries, we shall find that its geographical features have determined the position and limits of successive races which have settled within its area. For this county, with its adjunct the Isle of Ely, lies in the path of all movements of migration, trade, and invasion between the Thames basin on the one hand and the strongholds of the ancient British tribe, the Iceni, in the north and the shores of the Wash on the other. Along this strip of country, within the county's boundaries, we find the remnants of every race which has occupied the Eastern Counties, and relics of every civilisation and every invader, from the mammoth, rhinoceros, and hippopotamus, to the latest human settler.

No river or strong natural feature marks the southern boundary, but on the east and south-east lie the clay-covered chalk uplands which, when they were clothed with impenetrable brushwood and forest, were as good as a buffer state. They formed a tract of no-man's land, and it was only in later times when the woods were cleared away and men had encroached further and further upon the forest that arbitrary boundary lines were fixed by agreement, founded on local reasons of which we now know nothing.

Along the downs, between these wooded uplands and the fens, lay the great natural land-route to the thickly populated country on the shores of the North Sea. The early agricultural peoples who occupied this strip of land protected themselves from raids and encroachments by vast earthworks, the "dykes," which we shall mention later. They were constructed at intervals across the

Thorney Road, Flooded, Whittlesea

only open ground, and their great height and length are a wonder to us now in the present day.

On the north and north-west the swamps and dense forests of the fenlands afforded as sure protection against attack as did the tangle of woods on the south-east. Although terraces and spurs of gravel ran far out into the marshland, only natives acquainted with its intricacies could find the way in dry weather over land—in wet weather, when the floods were out, by canoe or boat—to Ely and the other isles. Whoever held this district commanded its slow-running waterways leading from the sea into the heart of the country. This northern district, known as the Isle of Ely, has a remarkable history of its own, which explains its position as a little county within the county. From early times it was famous for its abundance of fish and wild-fowl, and its islets, rising above the swamp, were fertile and "passing rich and plenteous," whilst the religious sanctuary on its highest ground afforded refuge to the inhabitants when hard pressed. Its first monastery, founded in the seventh century by the famous St Etheldreda, was swept away by the Danes; but under King Edgar, an abbey was built on its site and its revenues were restored.

So we see how Ely became important because of its natural surroundings and geographical position. The light of religion and learning was kept alive on the lonely island in the fens and all the ravaging of successive conquerors could not quench it. From this old beacon sprang the first spark of the University of Cambridge. The great Cathedral still stands, but only a few small patches of the original fen remain to tell us what the Isle was like in bygone days. As population and civilisation increased, efforts were made to reclaim the land, and from time to time various tracts were drained, so that a network of ditches or "lodes" now lead the water to pumping stations whose chimneys make familiar landmarks. Here steam-power lifts the water into larger channels which carry it to the sea, and now this area, which is in many parts below the sea level, has been almost completely reclaimed. The marsh has been drained, the forests burnt or cut down, and we find the physical geography, and consequently the climate of the district, completely changed by the agency of man.

The fen district is now, perhaps, the best-drained land in the country. It produces fine crops of wheat, roots, and vegetables, and the cultivation of fruit, which has been introduced both here and in other parts of the county, now forms one of its most important industries.

3. Size. Shape. Boundaries.

If some arbitrary administrator had divided up England, he might very likely have seen a convenient county in the basin of the Great Ouse ; the area drained by that river and its tributaries being about 2960 square miles, or some 349 square miles larger than Lincolnshire. But Cambridgeshire has not been formed in this way ; it has grown out of the conditions due to natural agencies and the political circumstances arising out of them, and it now consists of about 860 square miles, being more than a quarter of the area of the basin of the Great Ouse and less than a third of the size of the adjoining county of Lincoln. It ranks twenty-eighth among the counties of England in area. Cambridgeshire lies between 0° 16′ West longitude and 0° 31′ East longitude, its most westerly points being at Peterborough and near the railway at Tempsford station, and its most easterly near Kentford and Silverley. In a north and south direction

it extends from about 52° 1′ near Olmsled Hall, and
Odsey House near Ashwell in the south, to 52° 45′ near
Tydd St Mary's on the north.   It is thus roughly about
50 miles from north to south, and 30 miles from east
to west.   At Ely the breadth is not more than about
15 miles.

The meridian of Greenwich runs through the county,
passing about 4 miles west of Cambridge and just to the
east of Royston.   It has been more than once suggested,
in former years, that the astronomical work of the
Greenwich Observatory might with advantage be transferred from the smoky atmosphere of London, to the
clear air of the Royston hills, a new Observatory being
erected there on the Greenwich meridian.

In shape there is a certain resemblance between Cambridgeshire and the whole of Great Britain.   England, as
far north as the Humber and the Mersey, roughly suggests
the southern portion of the county up to Over and
Newmarket, whilst Scotland and the north of England
represent the fenland.

The counties which surround Cambridgeshire are on
the north Lincolnshire, on the north-west Northamptonshire, on the west Huntingdonshire, on the south-west
Bedfordshire, on the south Hertfordshire and Essex, and
on the east Suffolk and Norfolk.   For convenience of
administration, there have been of late years several
transferences of parishes, or parts of parishes, between
Cambridge and the adjoining counties.   Part of Papworth
St Agnes was taken from Huntingdonshire and added
to Cambridge, Royston was given to Herts, Great and

Little Chishall and Heydon were transferred to Cambridge from Essex, Newmarket and Wood Ditton from Cambridge to Suffolk.

The southern part, which has been based upon the original open ground between the ancient forest on the east and the fenland on the west (see Map, p. 133), is a broad oval with a very irregular outline, because its advance was not determined by physical features, but depended, among other circumstances, upon whether the folk on the Cambridge side, or those from the adjoining parts, first pushed their way into each portion of the uncultivated area between them. This southern division of the county has Cambridge as its chief town. The northern or fenland division, of which March is the central town, forms an irregular rectangle having Over and Isleham Fen as its southern, and Peterborough and Tydd St Mary's as its northern corners. It comprises the ancient division known as the "Isle of Ely," which, as we have before stated, now forms part of Cambridgeshire, though separated from the county for many administrative purposes—this independent recognition having arisen out of the authority in old times exercised by the Abbot of Ely. The boundaries of the Isle, defined, as Mr Conybeare tells us, by King Edgar in the tenth century, still hold good. "The eastern boundary," he says, "was marked by a still existing ditch, now called Bishop's Delf, whence the 'Isle' extended to the Nen at Peterborough, twenty-five miles in breadth, with a length (from 'the middle of Tydd bridge' to the Ouse near Upware) of twenty-eight miles." (Conybeare, *Hist. Camb.*)

The boundary of this northern division sometimes coincides with rivers such as the Catwater, one of the channels into which the Nene splits, near Peterborough ;

The Nene between Peterborough and Guyhirne from Dog-in-a-Doublet Toll Bridge

sometimes it follows artificial features, such as dykes, drains, and roads.   On the east the old river Nene, the Little Ouse, the old Welney river, and the Lark, approximately define the limit, while on the south-west and north-west it more commonly follows artificial water-courses. Ancient roads or earthworks mark the boundary for short distances, as in the case of the Icknield Way, east of Newmarket, or the Via Devana, and the Brent Ditch near Linton. This and the constant taking in or giving up of portions beyond the line, which was obviously adopted as a guide in fixing the boundaries, show that they are the result of arrangement and re-adjustment from time to time. Three tongues of Cambridge project into Bedfordshire on the south-west, and on the north-east the boundary both north and south of Wisbech runs now on the near, now on the farther side of the river. These irregularities may sometimes be explained in one way, sometimes in another; the changing course of a stream breaking into an old channel in flood time may have cut off what was at first part of one property and in one county. Rights of peat-cutting, felling timber, or digging gravel, may have developed into unopposed ownership, or, it may be, some long-armed baron or ecclesiastic may have claimed full proprietorship over areas where he had been accustomed to exercise rights of forestry or sport.

4. Surface and General Features.

A description of the surface of a county, as of a person, depends upon the nature and point of view of the observer. Cambridgeshire is like a quiet, reserved person who speaks little to strangers, but is full of surprises and wonderful stories for those who have learnt to love her with an intimate knowledge.   As we explore we shall find that the flat, dull, damp Cambridgeshire is only to be found in the mind of the tourist who pays a flying visit to Cambridge and Ely, and reads his newspaper, or goes to sleep, as the train whirls him through the fens.   It is the quiet villages, the chalk uplands with their carpet of fine grass and flowers, the lonely peat and fen lands—vast spaces of changing colour and mood, stretching away to the sunset—which are the real Cambridgeshire.

The best way to understand our county is to choose two or three high points, and from them make a survey of our own.   The geography which we learn like this, on the spot, remains with us when the map-in-school-knowledge has melted away like a dream.

First, then, let us choose a clear day and, starting from Cambridge, follow Hills Road on to the Gog-Magogs until we reach the high ground by Gog-Magog House.   We are now standing on the chalk ridge known as the East Anglian Heights.   The range sweeps across England from the Yorkshire coast, makes the cliffs at Hunstanton in Norfolk, crosses Suffolk and South Cambridgeshire, Hertfordshire, Bedfordshire, Buckinghamshire, and Oxfordshire, and extends as far as the coast of Devon.   At Nettlefold Hill, in Oxfordshire, it reaches 820 feet, and at Kemsworth Hill, near Wendover in Hertfordshire, 904 feet.   But the mean height of the range is between 400 and 500 feet, and the highest point of the Gog-Magogs near Cambridge is only just over 300 feet, though in the south-east of the county the ground rises to about 400 feet.   This is the greatest height above sea-level which we shall find in Cambridgeshire. To the native of a mountainous country our hills are mere rising ground, but everything is relative, and they are our Alps, which raise us above the plain into the purer air, with nothing higher, eastwards, between us and the Ural Mountains. Such must have been Camden's idea in 1607, for, though he had travelled in many parts of England, he writes: "Near unto Cambridge, on the south-east side, there appear aloft certaine high hilles, the students call them Gogmagog Hilles; Henry of Huntingdon termed them 'Amoenissima montana de Balsham,' that is, The most pleasant mountains of Balsham, by reason of a little village standing beneath them, wherein, as he writeth, the Danes left no kind of most savage cruelty unattempted."

Close by the spot which we have chosen on the hilltop, in the garden of Gog-Magog House, buried amongst beeches and ilex, lies Wandlebury—a pre-historic fort, with three ramparts of such shape as the Britons made—but nothing is known of it beyond its name. Southward we look upon a pleasant undulating land of pink plough and green downs, with lines of poplars and groups of fine timber, the long snake of steam from a Great Eastern train showing us the line of the valley of the Granta. On the south-west we see the Great Northern train which, nearer Cambridge, touches the valley of the sister stream, the Rhee or Cam, from Ashwell. From Baldock to Royston the railway follows the Icknield Way, that ancient route which ran, like the mid-rib of a leaf, all along the high ground from the south-west to the country of the Iceni, passing through Newmarket and Icklingham. Behind us, on the east, the raised grass-way, Worsted Street, commonly called the Roman Road, crosses it at right angles as do the other mysterious "dykes" at intervals, as we follow the uplands to the north-east. From the highest point of the so-called Roman Road

Source of the Cam at Ashwell

we overlook the chalk uplands, the scene of all the comings and goings and fights between the possessors and the invaders of our county.

Away on the west, across the valleys of the Cam, is the chalk ridge with its scanty mane of trees, the well-known landmark known as the "White Hill," rising between Haslingfield and Barrington, from which, on a clear day, 83 churches can be seen. Still further west, we have the plateau of older clays.

On the north, far away behind Cambridge, marked by the faint outline of Ely Cathedral, is the Fenland. Before we descend into these lowlands, which are in some parts but a few feet above sea-level, it will be well for us to realise that the high ground on which we stand is also a plain, levelled ages ago by the sea, and then raised and carved and moulded by rain and rivers, wind and weather. The hills are not pushed up as hills, but are the land left between the valleys which have been scooped out since the plain was first uplifted.

We may see for ourselves at Hunstanton how the sea is eating away the land and also arresting the further action of the streams which are carving out the valleys. The streams cannot act below the sea—so the result at last is a plain a little below sea-level. The flat-topped hills all round Cambridge are part of a plain formed in that way; the plain has been lifted up several hundred feet above sea-level, and ever since that uplift began, rain and rivers, sun and wind, have been at work producing the scenery we have before us, varied according to the character of the deposits worked upon. In other parts of East Anglia we know that the sea crept over the chalk and laid down sediments unlike any which we find here, and that during a long period great earth movements took place which altered the direction of the rivers and the face of the land. We have few traces of these deposits nearer than Sudbury

in Essex, and the later boulder clays which we find on
our hills, in our valleys, and in terraces and spurs along
our old river courses, carry on the story to much. later
times, and bring us to that mysterious period known as
the Great Ice Age, about which we shall have something
to say in a later chapter.

Our second bird's-eye view shall be from the top of
Ely Cathedral, that great minster standing on the spot
where Etheldreda built her monastery in the country of
the wild Girvii.

Here, on this main island of the fens, was the camp
of refuge from the Norman conquerors, safe because of
its surrounding swamps and tidal waters, and famous, no
doubt, long before we know anything of its history. We
are now on the chief island in the political division known
as the Isle of Ely.

As we climb the tower we have peeps here and there
down into the Norman aisles below and out on to the
roof and octagon, and from the boss on the top of
the lead roof, on a clear day, we can see the country
for 30 miles around. The first impression, in summer
time, is that we are looking down on a nest of houses
in a sea of golden corn. Southward along the line of
the Cam and the Great Eastern Railway, lies Cambridge,
marked by its spires. Haddenham's square tower, on the
wooded hill to the south-west, shows us where the Isle
dips down to Aldreth, where Kingsley, in *Hereward the
Wake*, tells how twice the Conqueror tried to enter by a
causeway of stones, trees, and hides, which he raised on
the marshes, and by a floating bridge thrown over the

Old West River. It was there, so the story goes, that Hereward and the English watched the Normans drown in the black mud as their bridge overturned, and there that the English set fire to the sedge and burnt the enemy out as they tried to cross by the causeway. Next, on the hill, stands the great tower of Sutton church, and the ridge ends at Mepal where the road crosses the "Bedford

Aldreth Causeway

Rivers," the two artificial canals which run from Earith to Denver Sluice. Farther to the north-west lies March in the centre of the fens, and beyond it, northward still, is Wisbech in a wooded belt, with the lowlands on either side, intersected by old river channels and dykes. Clumps of trees mark small islands, mostly ending in the Saxon "ea" or "ey," meaning island. Such are Manea, Stonea,

Southrey; and some of these stand on ancient gravel banks in which are found marine shells and bones of extinct animals left here long ago by the sea, before the peat was formed.   At Burnt Fen, near Littleport, large

Sutton Church

cockle shells, belonging to a much later inrush of the sea, are found in clay, beneath six feet of peat.   The Little Ouse, or Brandon river, bounds this fen on the north, and carries the eye out of our county to the dry sands and pine

woods of Norfolk.   Close by we can follow the old bend
of the Cam below Stuntney, which was once an island,
with the Bishop's Delf running east of it.   At Soham,
marked by its grand tower, was once a famous monastery,
the residence of the first bishop of these parts, St Felix,
who in the seventh century gave his name to Felixstowe.
Five hundred years later, the first Bishop of Ely made a
causeway across Soham Fen to connect Ely with Stuntney,
and this was the first road into the Isle, except William
the Conqueror's military roadway at Aldreth.   The present
road, which probably runs along the old way, leads us to
Newmarket and the chalk downs, where, beyond the
mills and pumping stations of Upware and the peat
cuttings of Wicken and Reach, we see the hills crossed
by the Icknield Way and the old dykes.

The road below us passing northwards through Ely
from Cambridge is said to follow the line of the Akeman
Street, which the Romans made across the fens.

The present river course, by the Cutter Inn, carries
us past wharves where in May and June we can still see
rows of women peeling the willow wands, for that most
ancient British trade of basket-making which is still carried
on here.   By the river, further on, we come to Roslyn Pit,
where a great quantity of the Kimeridge Clay has been
cut away to make and repair the river banks.   This is
the happy hunting-ground of naturalists by night and day,
and here is a puzzle for the geologist—a huge mass made
up of Greensand, Gault, and Chalk has been tumbled
into an ancient valley out of its proper place, probably in
the time when the ice passed over the land, twisting and

kneading the sea cliffs of Norfolk and Suffolk, and setting hard riddles for us to guess.

The tower of St Wendred's Church at March is another good point from which to realise the commanding position of Ely, the site of Whittlesea Mere, and the rich wheat-lands of the fens, with their numberless windmills

Peeling Willows

and gleaming waterways.   But the view of the roof alone, from the little gallery in the tower, is worth a long day's journey : the great carved beams and row upon row of hovering angels are doubtless the work of some beautiful mind, whose vision of the Heavenly Host remains to us in the dark fen oak.

5. The Fens.

From a bird's-eye view over different parts of our county we have gained some ideas about its most striking features; but a district so unusual as the Fenland needs a separate description. This low-lying country extends on either side of the Wash and reaches as far north as Lincoln.

Once upon a time the chalk rocks ran right across the Wash, forming a barrier from Hunstanton to Skegness. There were clays, as now, sloping gently under the chalk, and rivers, running over these deposits at the back of the barrier, formed plains and valleys on the south-west and cut a gorge through the chalk out to the sea somewhere opposite Hunstanton. That was the first stage. Then the ebbing and flowing tide and wind-lashed waves widened the estuary and ponded back the rivers, till beaches were formed along the seashore and mud banks in the rivers. That was the second stage. Sometime in the course of these changes the Ice Age set in, and when that had passed away, the sea still found its way far into the land-locked basin, now covered with boulder clay, leaving banks of gravel with sea shells and bones of hippopotamus and rhinoceros and elephant at March, Whittlesea (p. 27), Manea and elsewhere. That was the third stage. At one time England was joined to the Continent, and instead of the North Sea a great river ran out northwards in continuation of the Rhine, of which the Great Ouse was a tributary. We

Meeting of the Burwell and Reach Lodes, near Upware

know, by finding roughly-chipped flint implements in the old river gravels of this age elsewhere in the district, that man, belonging to the earliest stone age, once lived here; he hunted the big game, but we find no traces of him in the peat of the fens. Both he and the great beasts had passed away by the close of the third stage, before the peat, now seen in the fens, had been formed. We must remember that these changes occupied a vast number of years.

And now the history of the existing fenland begins. The land seems to have risen a little, and gradually most of the boulder clay, left by the ice, was carried away. The river mud and silt advanced as the sea fell back, and great dunes were formed by sand blown from the exposed shore. Behind these the fresh water was held back, peat was formed, and by and bye trees grew. There were still changes of level, for we find stumps of trees out beyond low-water mark on the present shore, but within the barrier of the silted-up estuary the fens grew uncontrolled. We find abundant remains of man of the later stone age. We shall hear in another chapter of his polished implements and of his hunting the great wild ox, also of his successors who used bronze and iron. All these early tribes found food and refuge here, but the Romans were the first people who had a plan for draining the fens. Tacitus says the Britons complained that the Romans wore out and consumed their bodies and lands in clearing the woods and banking the fens. They began the work of reclamation of the fens—that is of their destruction.

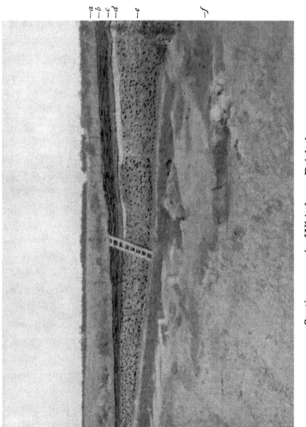

—a
—b
—c
—d

—e

—f

Section seen in Whittlesea Brickpit.

(a Surface soil; b Peat; c Peaty marl; d Shell marl; e Gravel (marine);
f Oxford clay)

But before man interfered with the fenland, natural changes produced wet ground here and dry there, at one time allowing the growth of trees, at another of nothing but swamp plants. Between Littleport and Ely, and in other parts, evidence of successive forests has been found. Huge oaks lie buried in the peat, some of them 90 feet long before throwing out a branch; also birch, Scotch fir, yew, and hazel. Changes have been going on all the time over the whole area, banks being destroyed in one place and heaped up in another, and we must remember that for a great part of the time this land was sinking, so that wet conditions prevailed more and more. The trees could not live in a swamp; they would decay as the peat grew and would be thrown down by some sudden storm, then as now, generally from the south-west, as their position shows. We can imagine the untouched fen; a vast plain covered in most parts with sedge, water plants, willow, alder, guelder rose and bog myrtle. Shallow meres, changing in size according to the season, were joined one to another by winding water-courses. Forests grew on the higher, drier parts and the islands were very luxuriant in summer time. All the swamps were accessible during hard winters, when the frost made everything solid. Then the fenmen no longer "stalked on high upon stilts" or used their long poles, but sped about on bone skates.

When we come to the natural history of the county we shall see by quotations from early writers what fine trees, fruits, and crops were grown in the fenland before 1200. Then things changed for the worse, and Dugdale's *History of Imbanking* tells us of the long struggle to cope

with silted-up water-courses, inrushing tides, and drowned
land, which began after the terrible inundation of the sea
of 1236, and continued all through the middle ages.
Acts of Parliament for draining the fens were passed during
the reigns of Elizabeth and James I, and in the reign of
Charles I a Dutch Engineer, Sir Cornelius Vermuyden,
undertook to drain the great level, for which he was

Ancient Bone Skates

to receive 95,000 acres of land. The fenmen however
were prejudiced against foreigners and would have none
of him. But though his contract was annulled his plans
were finally adopted and carried largely by the influence
of Francis, Earl of Bedford, a generous and true friend to
the fenmen, who with thirteen other gentlemen took up
the venture and in 1630 signed an agreement at Lynn,

known as the " Lynn Law." These were the original
" Adventurers" who gave a name to parts of the fen.
The work, which was interrupted by quarrels amongst
the landowners and the company, was then taken up by
the King. It was finally completed after the Civil War
in 1653 by a new company, still under the Lynn Law,
formed by William, Earl of Bedford, son of Francis
mentioned above, from whom the " Bedford Level"
takes its name. So Vermuyden's scheme was carried
out. The natural system of drainage was set aside—
rivers were altered and turned about; and new cuts were
made to the sea. The country was divided into three
areas, the North, Middle, and South Levels, crossed by
large straight drains into which smaller drains, guarded
by sluice doors, emptied themselves. Twenty years after
the drainage was completed, the peat had shrunk so much
that the side drains no longer discharged themselves into
the main channels. Windmills were erected to lift the
water up to the higher level, but now steam pumping
works have taken their place, and soon the picturesque old
mills will no longer—as an eighteenth century novelist
has it—"lend revolving animation to the scene."

A few patches of undrained fens are left in the valleys
by the rivers near Hauxton, Quy, Bottisham Lode, and
Chippenham.

Close to Cambridge, on an island on the river, by
Coe Fen and on Sheep's Green, in small ponds which
represent deserted parts of the river bed, rare shells,
plants and insects still linger—and long may this priceless
possession of the town remain unspoilt by road and traffic.

But the best known and most interesting relic of true fen is Wicken. Here we find the peace of the fens unbroken—the swish of the reeds against our barge and the gurgle of the water make a sleepy sound; ditches instead of hedges

Burwell Lode

cut up the plain; three or four black poplars rustle over a thatched cottage by the water side; here and there, in the distance, is a clump of willows, a pumping station—an old windmill; the peat-cutters or the mowers may be at work—barges full of peat and sedge are passing along

the lode towards the open river. The waterways divide,
a smaller lode turns to the sedge fen, where the reeds

Peat digging, Burwell Fen

grow breast high along the water side, guarding from
sight the last home of rare and beautiful things, which,
almost everywhere else, have been destroyed by man.

Peat at Burwell Fen

H. C.

3

A bicyclist can ride from Burwell by the chemical works, across peaty fields crossing one or two ditches on planks and lifting his machine over a gate. Chippenham Fen, approached by an ancient drove bordered by Scotch firs, is more remote and is even more interesting than Wicken.

Peat is dug everywhere for local use, but the largest diggings are those in Burwell Fen, from which it is transported in barges along the lodes and rivers for distribution to more distant places. The peat, which is always called turf by the fen people, is sometimes dug in trenches a yard wide and from five to six yards apart, the depth depending upon the level at which the water stands in them. These trenches get filled up in time with sludge, oozing from the surrounding mass. Sometimes the whole is removed, so that a wall of peat stands in front of the work. A special flat spade is used, shod with iron, with a small flange, formed by turning up the side, so as to cut off a uniform brick-shaped block. A thousand blocks should weigh a ton and, as fuel, should be equivalent to about half a ton of coal.

The upper part of the peat consists of a spongy mass, in which the roots and leaves of which it is made can be still distinguished, but in the lower part, where it has undergone further decomposition and pressure, it is a more compact black mass. The moss which has contributed most to the formation of the fen peat is *Hypnum*, whereas the hill peat of northern counties is made up chiefly of *Sphagnum*, which does not occur here.

6. Watershed and Rivers.

A glance at the map on the next page will explain many points of importance in the geography of Cambridge. In the first place it will be seen that the county covers an area less than one-third the size of the basin of the Great Ouse, also that the county boundary rarely coincides with anything that could be called a watershed.

The Ouse collects its waters from the surface of large areas of clay in the south-west, outside the county, and its tributaries, the various branches of the Cam, are fed chiefly by springs coming out from the receding chalk escarpment. On the flat-topped hills of the eastern and southern borders of the county the sources sometimes overlap and the watershed is very irregular.

About half-way between Saffron Walden and Bishop's Stortford, near the watershed between the Thames and the Great Ouse, the Cam, here known as the Granta, first takes shape as a distinct stream. It runs over various more or less clayey surface-deposits on the Upper Chalk and when it has reached the saturation level, it flows between sharp-cut banks through an alluvial flat by Littlebury, Chesterford, Ickleton, Hinxton, Duxford, and Whittlesford; leaving Pampisford and Sawston to the east. Here we see that it has reached the low-level lands, not much above the fens; it divides into several channels to reunite again a short way below. It is a sluggish stream and can only just hand on the material brought down from the upper part of its course

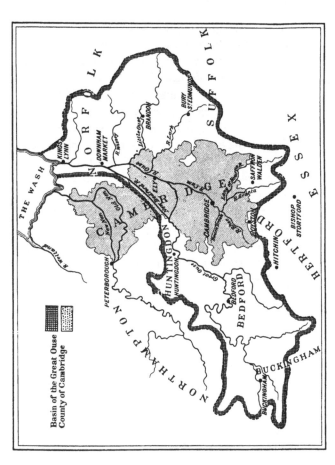

Map showing Basin of the Great Ouse as compared with Cambridgeshire

or washed in from the slopes on either side, and can rarely transport large masses of gravel, except when some accident has narrowed its channel or some other cause has increased the velocity of the stream.

We learn from the terraces of sand, loam, and gravel, left at various heights along the valleys, far above the present water-courses, that the rivers have been at work for ages, long before they cut their way down to their present lower channels. These gravel terraces are of extreme interest, because they record the great changes which have gone on in our district both in its physical geography and in its plant and animal life. They are the most important geographical feature in this part of the county.

The Bourn River collects the waters of a number of small streams near Bartlow and runs by Hildersham, Abington, and Babraham through the rich, well-wooded, low ground which we saw when we looked south from the Gog-Magogs. Between Stapleford and Shelford it joins the Cam. This is not the same as the Bourn brook which we shall come to presently, running in from the other side.

The Cam and all its tributaries have, so far, drawn their waters either from the top of the drift-covered plateau, which forms such an important feature in East Anglian geography, or from the valleys which run far up into it. Now however the Cam leaves the chalk hills and their plain of marine denudation and enters upon a country of clays which underlie all the fenland.

The branch of the Cam known as the Rhee rises at

Ashwell in the north-east corner of Herts. half a mile over our border. In a hollow, shaded by ash trees and reached by a little path from the village street, the clear water gushes up out of the chalk, in many springs, and

Old Bridge near Haslingfield

makes at once a stream of ten yards wide. Few rivers have a more beautiful source. It flows by Meldreth, Barrington, Harston, and Haslingfield, gathering water from the clayey ground on the south and west, and from the base of that spur of chalk called the White Hill, which lies on its left bank. The beautiful field path from Harston leads past Burnt Mill bridges to Cambridge. East of Cantaloup Farm the Rhee joins the Granta and it may be followed down to Lingay Fen and Byron's Pool where the Bourn brook[1] joins the main river.

The Cam from King's College Bridge

At Byron's Pool, where the poet is said to have practised his swimming feats, flood gates now turn part of the water off to the present mill. Somewhere

[1] The Bourn Brook, sometimes also called after its largest tributary the Long Brook, has its rise in the Eltisley fields beyond the village of Bourn; it collects the waters off thirty or forty square miles of Gault in the west and taps a wide area of chalk. It must not be confounded with Bourn River by Abington and Babraham. This or Bourne is a common river name and probably is merely a form of the Scotch burn.

near here stood Chaucer's Mill on the bridge "At Trompyngton not far fro Grantabridge" made famous in the Reeve's Tale.

The upper river has had its tragedies and no undergraduates are allowed upon it until they have passed a strict swimming test. In the seventeenth century, scholars were punished for bathing anywhere in Cambridgeshire

The Cam at St John's College

by a flogging, whilst Bachelors of Arts, for the same offence, were made to sit all day in the stocks in their college hall.

The Cam—happily it has but one name now—runs through Cambridge, past the old mills and "the Backs," where its ancient course may be seen, now here, now there, winding about from one side to the other, repre-

sented by ditches which have been modified so as to form
in part the boundaries of the College grounds. In early
times all this was marsh land and was covered with a dense
growth of trees: it is called in the old Field Books
" Thousand Willows." The old river channel was once
far below the bottom of the present river, for when trial
holes were sunk in the Trinity paddocks, the river silt
was found to rest on the Gault, in one place at a depth of
forty-five feet. That is sixteen feet below ordnance
datum (mean sea-level), which proves that a depression to
that extent has taken place since this channel was made.
The present level of the river at any part must not be
taken to represent what it was in old times before the
water was held back by locks and weirs.

Between Chesterton and Baitsbite all the University
boat-races take place, the single and double sculls, pairs,
fours, and the bumping races of the College eights in the
Lent and May Terms—and Town races later in the
summer time. A sailing club, too, has many enthusiastic
members who know the intricate waterways of the
fenland.

From Cambridge the river is navigable, though it has
nothing like the water traffic of former days. It flows
northwards by Waterbeach and Upware and, after a course
of about forty miles, joins the Old West River, still called
the Great Ouse, and takes its name, three miles south of
Ely, at Thetford, a small village not to be confounded
with the Norfolk Thetford on the Little Ouse.

Ouse is an old river name meaning water. We find
it in Buckinghamshire, Sussex, and Yorkshire as well as

The Lent Races. A Bump

in the foreign Oise and Auser. The Great Ouse is the most important of the Cambridgeshire rivers, and we have seen that its basin extends far beyond the county on the east and west. It rises on the borders of Northamptonshire and Oxfordshire and flows through Buckingham,

The Cam at Upware

Bedford, St Neots, Huntingdon, and St Ives, entering the fens at Earith. Not far from Earith is Overcourt, with its old inn and ferry; one of the most charming spots on the Ouse. Formerly the river here sent a branch to the Nene and another to the Old West River which channel still runs round the Island of Ely, by Aldreth, to join the Cam, but it is now reduced to a small stream. As was natural on low land with an outfall which was being continually silted up, the fenland rivers have altered their course many times, but the systematic drainage and division of the fens into three areas, the North, Middle, and South Levels, each independent of the other, has entirely changed many of the old water-courses and given the rivers artificial channels to the sea.

After passing Ely and receiving the Lark and the Linnet, the Ouse is now diverted near Littleport towards the Little Ouse; it leaves Cambridgeshire and is then carried past Denver to the sea at Lynn—instead of following, as formerly, the Old Croft or Welney River to Wisbech. But the great change made in this river was when the Middle Level was drained, in the seventeenth century. The Ouse was embanked and brought from Over to Earith, where a sluice was made and the greater part of the water was turned into the New Bedford River, one hundred feet wide and twenty miles long, which was cut from Earith to Denver, parallel to the Old Bedford River, so as to convey the high-land waters more directly to the sea at Lynn. A bank on the outer side of each river separated the Middle from the North and South Levels and the "Wash," or space between the two rivers, was "for the waters in flood times to bed in." The bottom of the New River was eight feet higher than that of the Ouse, and a sluice was first made at Denver in 1653 against the tidal waters of the north, to prevent the Bedford Rivers flowing back up the Ouse.   Before this sluice was made the tide flowed up the Cam to within ten miles of Cambridge.   It now runs up the Bedford River to Mepal and sometimes to Earith.

The Old Bedford River

In 1713, during a very high tide and heavy river-flood, the sluice blew up and the Bedford River and the tide flowed up the Ouse into the Cam—so that all the South Level lands were flooded and much discomfort and quarrelling were occasioned between the landowners of the North and South Levels.   After thirty-seven years the sluice was restored and a new massive stone dam was built. Eventually a cut was made to straighten the river at Lynn and this allowed the silted-up channel to be cleared out by the tide.

The northern part of the county is drained by the river Nene, which rises in Northamptonshire and enters the fens at Peterborough. The main river now flows

The Nene at Foul Anchor

through an artificial cut, known as the New Leam, nearly parallel to the older cut made by Bishop Morton in the fifteenth century. It passes north of Whittlesea, through Guyhirne and Wisbech, and leaving our county enters the Wash twelve miles north of Wisbech. In winter the low land between the two channels, known as Whittlesea Wash, is still often under water. But before the fens were drained terrible damage was caused by the giving way of the banks during floods and by high tides. The church bells were at such times rung backwards to warn the people, and the flood often came so suddenly that the farmers had not time to drive their cattle to a place of safety, but were glad to escape in boats before their cottages were swept away by the wild rush of water.

Wisbech at High Tide

A second branch of the river is the Muscat or Catwater, north of Peterborough, known also as the South Ea, or Shire Drain, which is now carried into the estuary by the North Level main drain at Tydd Gote. The third branch, or Old Nene, flows by Whittlesea and Benwick to March and to the Welney at Salter's Lode Sluice.

All through its history the town of Wisbech has had great difficulties to contend with from the tidal waters, until 1830, when the new Nene Outfall Cut to Crab Hole was devised by two well-known engineers, Rennie and Telford, and was carried out at a cost of over £200,000.

Wisbech at Low Tide

The river was diverted from its old channel amongst the shifting sands. A Wisbech historian relates how " The whole river, from Wisbech to the sea, was now wound into nearly a straight direction " with one interesting result that the tidal wave, or " bore," deserted the river on the opening of this new outfall. The windings and obstructions formerly held back the inrushing high tides, until the accumulated water forced itself in a great wave along the narrow passes of the river. In 1680 Ralph Thoresby described it in his diary: "This morning," says he, "before we left Wisbech, I had the sight of an Hygre or Eager, a most terrible flush of water, that came up the river with such violence that it sunk a coal vessel in the town, and such a terrible noise that all the dogs in it did snarl and bite at the rolling waves, as though they would swallow up the river, the sight of which (having never seen the like before) much affected me. Each wave surmounting the other with extraordinary violence!"

7 a. Geology and Soil.

When the Geological Survey began to map England it was necessary to distinguish between the older beds which make up the earth's crust, such as limestones, sandstones, and slate, and the newer deposits which have been spread over the surface of the land in comparatively recent times, such as boulder-clay brought by the ice, sand, gravel, loam and silt left by rivers and tides. These latter they spoke of as "superficial deposits." The older deposits beneath these they called "rocks," whether they were sand as at Potton, or clay, like the Gault, and they all came under the head of "Solid Geology."

Our map shows the solid geology of Cambridgeshire, and also some of the more recent deposits of the rivers

DIAGRAM SECTION FROM SNOWDON TO HARWICH, ABOUT 200 MILES.

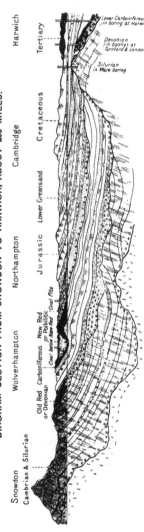

Snowdon

Cambrian & Silurian

Wolverhampton

Old Red Carboniferous New Red
or Devonian   or Poikilitic
                Coal below New Red   Coal Pits

Northampton

Jurassic   Lower Greensand

Cambridge

Cretaceous

Harwich

Tertiary

Lower Carboniferous
in boring at Harwich

Devonian
in borings at
Turnford & Lonson

Silurian
in Ware boring

This cross section shows what would be seen in a deep cutting nearly E. and W. across England and Wales. It shows also how, in consequence of the folding of the strata and the cutting off of the uplifted parts, old rocks which should be tens of thousands of feet down are found in borings in East Anglia only 1000 feet or so below the surface.[1]

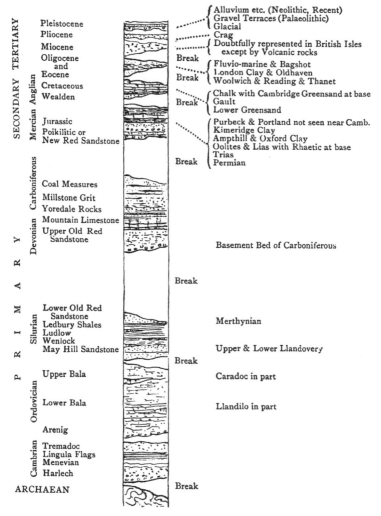

SECONDARY TERTIARY	Mercian Anglian	Pleistocene		{Alluvium etc. (Neolithic, Recent) Gravel Terraces (Palaeolithic) Glacial
		Pliocene		Crag
		Miocene		{Doubtfully represented in British Isles except by Volcanic rocks
		Oligocene and	Break	
		Eocene		{Fluvio-marine & Bagshot London Clay & Oldhaven
			Break	Woolwich & Reading & Thanet
		Cretaceous		
		Wealden		{Chalk with Cambridge Greensand at base Gault
			Break	Lower Greensand

Pleistocene
Pliocene
Miocene
Oligocene and
Eocene
Cretaceous
Wealden

Jurassic
Poikilitic or
New Red Sandstone

Coal Measures
Millstone Grit
Yoredale Rocks
Mountain Limestone
Upper Old Red Sandstone

Lower Old Red Sandstone
Ledbury Shales
Ludlow
Wenlock
May Hill Sandstone

Upper Bala

Lower Bala

Arenig

Tremadoc
Lingula Flags
Menevian
Harlech

ARCHAEAN

Alluvium etc. (Neolithic, Recent)
Gravel Terraces (Palaeolithic)
Glacial
Crag
Doubtfully represented in British Isles except by Volcanic rocks
Break
Fluvio-marine & Bagshot
London Clay & Oldhaven
Break
Woolwich & Reading & Thanet
Chalk with Cambridge Greensand at base
Gault
Break
Lower Greensand
Purbeck & Portland not seen near Camb.
Kimeridge Clay
Ampthill & Oxford Clay
Oolites & Lias with Rhaetic at base
Trias
Break
Permian

Basement Bed of Carboniferous

Break

Merthynian

Upper & Lower Llandovery
Break

Caradoc in part

Llandilo in part

Break

Break

SECONDARY TERTIARY
Mercian Anglian

PRIMARY

Carboniferous
Devonian
Silurian
Ordovician
Cambrian

This vertical section is on the scale of about 30,000 feet to the inch, and is intended to show the order of succession of the rocks in the crust of the earth and to indicate the horizons at which we have evidence in the British Isles of important interruptions in the continuity of deposition—that is to say, periods when the land was above water and was being worn away. The material thus washed away must have been laid down elsewhere, and therefore there are somewhere deposits which come in where breaks are marked in this section. On such a small scale it is impossible to put in many subdivisions, or to preserve strict proportions.

and fens, marked in buff colour, but the glacial deposits are not indicated. Two groups of formations are shown on it trending—or to use the technical term, "striking"—across the country from south-west to north-east.

The group on the north-west is coloured pink and drab, and that on the south-east various shades of green and pale grey. If we walk across these beds from south-east to north-west we shall find them coming out one from below the other in descending order as shown in the following section :

Diagram Section across Cambridgeshire

- (a) Chalk (grey on map).
- (b) Gault (dark green).
- (c) Lower Greensand (light green).
- (*) Unconformity.
- (d) Kimeridge Clay (pink).
- (e) Ampthill Clay, Corallian, and Elsworth Rock (red and lightest green).
- (f) Oxford Clay (drab).

We may here and there find a cliff or a pit in which the upper or newer beds can be actually seen resting upon the lower or older beds. We shall learn to know the different beds which make up the grey and green or the pink and drab groups, and we shall find that each bed belonging to one group has everywhere the same bed below it, until we come to the *bottom bed* of the green group, which rests sometimes on one and sometimes on another bed of the pink and drab group.

This lower group consists chiefly of stiff clay of a lead blue colour, with here and there beds of impure limestone, which never extend very far continuously. They were deposited in a not very deep sea, the shore of which was nowhere near our part of the country. The lower beds of the group do not come to the surface anywhere in this district, and the upper beds are hidden beneath the overlying group, so that we see neither the top nor the bottom beds. The whole group is called Jurassic from its being well seen in the Jura Mountains. Only three of its newest divisions are seen in our country. They have been divided and named according to their composition, their texture, and the fossils in them.

The first we take is the lowest of those which are found here. It occurs about Oxford and has been called the Oxford Clay, but it is best seen in the great brick pits at Peterborough and Whittlesea. This deposit was laid down on a sinking sea bottom until it reached a thickness of 500 feet. There were no banks of sand or gravel—nothing but fine mud—on which lived innumerable sea reptiles, fish, and shell-fish, whose remains helped to supply lime for the limestones. The sea bottom went on sinking about 200 feet more, and more clayey sediment was laid down on the top of the Oxford Clay, and a somewhat different set of creatures began to migrate into this area. This deposit was named the Ampthill

Clay from the place where it is best seen. In many places layers of calcareous mud were formed where groups of shell-fish and animals had crowded together or where their dead shells had drifted; so we generally find lens-shaped masses of limestone associated with this middle division of the older group. They can be seen at Elsworth and just above the great clay-pits at St Ives, and these are called Elsworth Rock. At Upware we have a limestone made up largely of corals as well as of sea-urchins and other animals. This was probably the débris of a coral reef, with the remains of innumerable creatures that lived among the corals.

Still the sea bed sank and another great deposit of clay was formed, which is named after a village in Dorsetshire, the Kimeridge Clay. This is well seen in Roslyn Pit near Ely, where it is largely dug for repairing the river banks. In the Kimeridge Clay we find that new and different kinds of creatures have come in and taken the place of the Oxford Clay species which had died out, or moved away.

It is clear that if these rocks were still lying flat as they were originally laid down we should see nothing of any of them except the top bed, unless we dug a hole some 800 or 900 feet deep. But we find them coming out one from below the other, as we travel in a north-westerly direction, which shows that they have all been tilted up on that side and their edges cut off, so that we can count and measure them across the country. One of these great movements of the crust of the earth took place and went on for a long time after the Jurassic

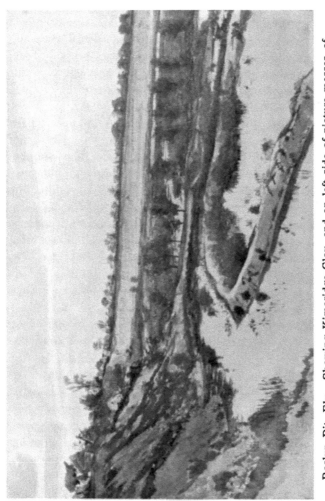

Roslyn Pit, Ely. Showing Kimeridge Clay, and on left side of picture masses of Boulder Clay with Greensand, Gault and Chalk, dropped there by ice

Rocks, which have just been described, had been deposited, and before the first of the next group, called Cretaceous, and coloured green and grey on the map, was laid down. The clearest proof of this is that the bottom bed of the Cretaceous group rests upon different parts of the Jurassic. It has been pointed out that we could not see the lower beds unless they had been tilted and the upper parts washed away; so the base of the Cretaceous could not have been laid down on lower beds of the Jurassic unless the upper beds had been previously removed. Thus we know for certain that the Jurassic Rocks were tilted up and cut away by sea and rain and rivers, and then depressed again below the sea, before the Lower Greensand was deposited. All this is briefly stated in technical language by saying that the Cretaceous Rocks are here *unconformable* upon the Jurassic, or that there is an *unconformity* between them. This unconformity is well seen at Sandy and at Gamlingay.

We will now try to read the history of the upper of the two groups of rocks which prevail in Cambridgeshire. It is called Cretaceous, from *creta*, chalk, because the greater part of it consists of chalk, though the lower part is clay and sand.

At the base we find a deposit of from 70 to 150 feet of sand with beds of gravel in it. The sand and gravel are quite loose and incoherent except where the infiltering surface water has rusted the iron which pervades the mass, and thus produced a cement which converts the loose sand into a gingerbread-coloured sandstone. This formation is called the Lower Greensand, to distinguish it

from another somewhat similar sand which in the south
of England occurs at a higher horizon and is known as
the Upper Greensand. "Why *green* sand?" may well be
asked, for wherever we see it, it is of a yellow or rusty
red colour. But it is so called because where it has not
been acted upon by the water soaking in from the surface,
it is found to be of a dark sage green, due to little films
of a green silicate of iron which coat each grain. We
generally see it where it has been exposed to the action
of the weather; but where it is reached in wells, where
overlying impervious beds have protected it from the
surface water, it retains its green colour. Where sufficiently
large masses have been hardened by the rusty cement, the
rock is known as Carstone, and is used for building. (See
p. 113.)

The gravel beds in it are of greatest interest, because
they are made up of lumps of the underlying clays which
have become hardened by phosphate of lime. Fossils
also out of the older rocks have been washed out and
similarly phosphatised. All this took place when, as has
been explained above, the older rocks had been tilted up,
and their edges were being washed away. The fossils are
generally much rolled and worn, but are still recognisable.
About 75 per cent. of them are fragments of a well
known Kimeridge Clay ammonite known as *Ammonites
biplex*, from the way in which the ribs on its surface
branch into two. These phosphatised lumps are of value
in agriculture, so we shall have to speak of them again.
Later there was a sudden and complete change in the
character of the sediment, and instead of sand there was

fine mud deposited to a depth of some 125 feet. Of course we should not expect to find the same creatures living on a muddy sea floor as those which lived among the strong cross currents which transported the sand we

Ammonites biplex

have been describing. This mud is called the Gault, a word which means clay. It is of great commercial value, as it is the clay of which the Cambridge bricks are made. It is also an important formation because it forms an absolutely impervious layer between our two water-bearing strata, and enables us to obtain water by the Artesian system. Still the sea bottom kept on sinking, perhaps not quite continuously, but, as there is some reason for believing, with stationary periods now and then. Less and less sediment reached this area and when it began to be scarce, and there was abundance of animal life, including fish and saurians, whose decaying bones furnished phosphate of lime, a layer from eight to fourteen inches thick, of chalky mud was formed full of lumps of phosphate and phosphatised fossils.

A little mud was still carried into this area, so that about 70 feet of the base of the chalk is so clayey that it is called Chalk Marl. The depression went on however till there was practically no mud carried into this area, and the sea bottom was raised only by the shells and bones of animals, and thus that great formation, the Chalk, was laid down. Much of it is made up of the remains of very small creatures, myriads of which floated in the clear water, and when they died their tiny shells settled down and helped to build hundreds of feet of chalk. The chalk was not a deep-sea deposit, and as we might expect from the manner in which it was formed, is of very uniform character all through. Yet in the long time required to build up 600 feet with only fragments of bone and cast-off shells, changes did take place which made the area suitable for different kinds of creatures, and the chalk has been divided up into zones according to the remains of different forms of life which characterise them.

In the upper part of the chalk the silica which some of the creatures used for building their skeleton, or their shell, such as the spicules of sponges or the siliceous parts of foraminifera, was set free and replaced portions of the chalk by flints, which occur sometimes in layers and sometimes lining joints. These flints—which could not be

The Chalk Quarry, Norman Cement Works

dissolved and got washed out when the chalk was wasted away—now cover the country, furnishing the material or many of the newer deposits. Their commercial value is great in a country like this, where there is so little stone.

Owing to their more clayey nature the lowest beds of the chalk known as the Chalk Marl behave with the Gault, throwing out springs and running far out over the low ground, while the main mass of the chalk rises in bold hills from the plain, and forms the beautiful undulating downs such as we see on the race-courses of Newmarket, or the training ground and golf links of Royston.

Chalk is largely quarried wherever it occurs, but the most valued beds are near the base. The very lowest part is seen in the pit at the Saxon and Norman Cement Works, and the next succeeding beds are exposed in the great and conspicuous quarries at Cherryhinton, while the chalk with flints can be well studied in the large quarries at Balsham.

7 b. Geology and Soil.

If we were to walk out west from Cambridge along the road to St Neots, we should expect from the above descriptions to see beds belonging to the Jurassic group wherever there were rocks exposed; and similarly, if we walked east over the high flat-topped hills, towards Dullingham for instance, we should expect to see chalk everywhere. But, instead, we should find the ground covered—to a depth much greater than anything touched by ordinary agricultural operations—with a stiff lead-coloured clay full of stones of great variety and sizes derived from rocks which have been recognised in places all along the east of England and Scotland, and even in Norway and Sweden.

This formation is not shown on maps of the "solid geology," and to follow its distribution we must procure another survey map, called the "Drift Map," on which this boulder clay, as it is called from the boulders or stones in it, is indicated by a lilac-grey colour. We shall find on this map also long strips of a reddish colour intended to represent ancient gravels left at various times along the valleys. Now what is this boulder clay? It is not so very long ago that it was ascribed to a great rush of water which, it was supposed, carried the big stones and clay and dumped them down higgledy-piggledy where we see them. There were, however, many difficulties in the way of accepting this explanation. A flood ought to have sorted the material differently. The stones were grooved and scratched in a way which showed that they had not been carried by water, as in the figure on the next page of one found in the boulder clay at Roslyn Pit. At last it was recognised that the boulder clay resembled the material transported by ice, and now these deposits are called "glacial" from *glacies*, ice.

Many theories have been proposed to explain why this great cold prevailed and passed away, but they are too complicated to discuss here. What we do know is that over all this part of England there once prevailed for a time climatic conditions, which not only interrupted the continuity of life—both of plants and animals—but also left such a mark upon the geographical features and land surface as has affected all the later denudation and determined the distribution of man and his cultivation of the soil. This episode is spoken of generally as the glacial epoch. The land was invaded by ice. Some say that it came in the form of a great ice sheet across the North Sea from Scandinavia, passing along our northern coasts. Some hold that the land was at a lower level, and that

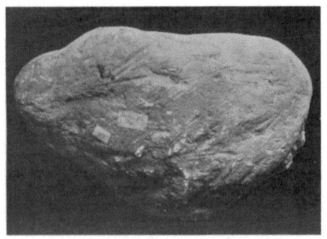

Ice-scratched Boulder of Chalk, Roslyn Pit

the ice was floated over it and stranded on it—as shore ice, pack-ice, or icebergs—but, whether by this or that means, the ground after it had been given nearly its present shape was covered by clay and stones transported by ice from far north, and even filling up ancient valleys, some of which have been nearly cleared out again, some of which remain filled. Those who hold the floating-ice theory point out gravels, etc., due to the winnowing of the glacial material while it was being dropped and sorted over the submerged land, and gravels due to the shore wash, as the sea bed was being raised. Those who favour the land-ice theory see in these gravels and sands material pushed up by the ice from the sea bottom, or the result of the streams that flowed in and from the ice sheet. After a time similar conditions to those which prevailed before the great ice age returned. Rain and rivers again washed down the surface soils, and heaped up the débris here and there where their velocity was checked. To these deposits, then, we turn with great interest to find out whether there is anything buried in them which will tell us what happened just before our own time.

There is near Barrington a bed of gravel and loam containing rolled fragments of rocks, which do not occur in this drainage area and which we therefore infer must have been transported over the watershed by the ice and then got washed out of the boulder clay into this gravel. In it we find the remains of hippopotamus, of an elephant whose teeth and tusks and bones are like those of the African elephant, and of a rhinoceros which is elsewhere found associated with creatures which are supposed to have liked a warm climate. With these are bones of the lion, hyaena, bear, bison, and Irish elk. Many of the bones are gnawed, as we usually find to be the case among the remains left by carnivorous beasts, especially by the hyaena. There are a few shells, apparently washed off the land or out of small ponds. These Barrington gravels

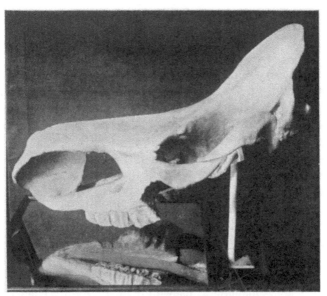

Head of *Rhinoceros Merckii*

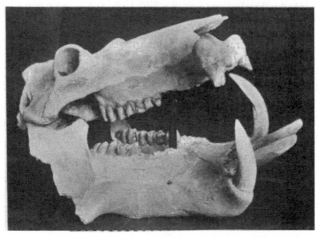

Head of *Hippopotamus amphibius*
(Both found in Gravel at Barrington)

thus seem to indicate a warm climate, but not far off—at Grantchester, Barnwell, and Teversham—we find other gravels in which are the teeth, tusks, and bones of another kind of elephant, resembling the mammoth now found frozen in Siberia, which had long hair and wool; and with it is a rhinoceros whose bones are like those of the woolly rhinoceros found in the frozen soil of North America. Among the shells of the last-named gravels are six kinds which no longer live in Britain. Four of them are shown on p. 67. The *Unio litoralis* (A), a fresh-water mussel, and another variety found here, live no nearer than the rivers of France. The other small fresh-water shell called *Corbicula* (*Cyrena*) *fluminalis* (D), which was found most plentifully in the Barnwell gravels, often with both valves still together, has gone to the rivers of Sicily, Asia Minor, and the Nile. This shell had lived in England longer than the others, and has now gone farther away than any of them. It and the Unio are two of the most characteristic of our gravel shells. The small spiral land shell (C) seen in the picture, is *Clausilia pumila*, now ranging from Sweden to Croatia, while the helix (*Fruticicola fruticum*) (B) lives in Northern and Central Europe. None of these are found now in the British Isles. Some shells found in the gravels are still in England, though not in this district, while others now common here are absent in the gravel.

There is one word much used in geology of which we in Cambridgeshire ought to know the meaning. It is *alluvium*. It means deposits of mud, sand, and gravel made in rivers and estuaries. We have in this county,

Extinct Shells from Cambridge Gravel

(A) A fresh-water mussel (*Unio litoralis*)
(B) *Fruticicola fruticum*
(C) *Clausilia pumila*
(D) *Corbicula fluminalis*

as we have seen, terraces and enormous spreads of gravel of exceptional interest. We ought to know how these terraces have been formed. When a terrace is 10 or 20 feet above the present alluvium, it gives the idea that the river has cut straight down from the higher to the lower horizon and that the height of the terrace is the measure of its age, but we must interpret the work of the river quite differently. The river cuts *back*, not *down*, and the measure of the age of the terrace, relatively to the present river level at any point, is the time that it has taken to shift back the rapids or waterfall, where the work is really being done, while the upper and the lower terraces are being formed.

Still the upper terraces are the older, and the time that elapsed during their formation was so long, and the geographical conditions of the south and east of England which affected the migration of the great mammals and also of the small water shells were such, that by the time we have got to what we have defined as alluvium, most of them had changed, and we need not wonder at this when we see how they are changing still.

In the terrace gravels and ever since their date, we find traces of man's sojourn here, and we see that his more ancient remains can be classified under two heads, the Palaeolithic and Neolithic, which are named after the older and newer type of stone implements he used, which will be described when we enter into the archaeology of the district.

8. Natural History.

The natural history as well as the political history of a district depends largely upon its geographical conditions and climate. Cambridgeshire has a varied and unusual combination of features—dry chalk uplands, clayey wooded tracts, meadow land, fenland, and tidal mud flats; so that plants and animals of very different habits find a home here.

The great mammals which are found so plentifully in the gravels of Cambridgeshire migrated here, as we have seen, in the time when England was joined to the Continent. The present English Channel is so shallow that if St Paul's Cathedral were placed in it anywhere, the dome would not be submerged. So no great change of level was required to join England to France. The animals travelled along the great rivers towards the north in summer and the south in winter-time, and seeds of plants were spread in various ways. Some of them had not reached Ireland before it was cut off by the sea. Many forms died out when they were no longer re-plenished from the Continent and many animals and plants are limited to definite areas—some have become extinct quite recently, and some of the butterflies and birds are rare visitors.

A short account of the gravels with the great beasts and shells found in them has been given in the chapter on geology, but the ancient fen fauna is linked more closely to our own times, and will perhaps be best described here.

With the remains of man, in the peat and alluvium of the fens, we find bones of the great wild ox, or Urus, which had lived on from the gravel days and survived until recent times on the Continent, though it had disappeared in Britain before the Romans came. There are bones too of the boar, the brown bear, and the wolf. When, exactly, the brown bear ceased to live in the fens we do not know; the only representative of his kind to be found here now is the badger, which is said still to breed in the west of the county. The wolf must have been common and have lived on until late times, but its remains are very rare and the fox is its only modern representative. The red deer remains are numerous and those of roe deer are also found occasionally. Both of these animals have migrated to the safer districts of the north and west of Britain. The fallow deer were never wild here—they were introduced, and only one herd, at Chippenham Park, now remains. The small Celtic short-horn ox, which lived on into historic times, has probably handed on modified traits to most of our breeds. That most beautiful and graceful of all our wild creatures, the otter, is still quite common in all our streams, though from its coming out only by night and from its silent glide into the water when disturbed, it is seldom seen, yet it travels along the rivers through Cambridge and has even been known to breed in the Backs of the Colleges. The marten or "sweet mart," and polecat or "foul mart," are gone; the stoat and the weasel, in spite of game-keepers, still help to keep down rats and mice.

The rodents, or gnawing animals, are numerous

here. The most interesting of them, the beaver, is gone, but its bones are quite commonly found in the fens, and it is mentioned as living in Wales in the eleventh century.

The brown rat has become a source of great loss from its voracious appetite and skilful burrowing. For some unexplained reason multitudes of these rats have recently appeared in this and the adjoining counties. More than 5000 were lately killed on a single farm in one season. When they came the old English black rat reappeared here and there. Several of these were seen by the writers lying dead by stacks above Linton. The pretty little water rat or vole belongs to this group. There are many kinds of mice and among them the dormouse and the little harvest mouse. Hares were enormously abundant before the passing of the Ground Game Act, but are now much reduced in numbers. Several species of bats are common, one extremely rare species (*Myotis myotis*) was caught alive at Girton some years ago.

From the earliest times Cambridgeshire has been rich in bird life. The early writers speak of the numberless waterfowl of the fens which in the twelfth century were so cheap that " for one half-penny and under, five men at the least may not only eat to stave off hunger and content nature, but also feed their fill." But the drainage and cultivation of the land have spoilt their haunts and exterminated many species. The bones of a young pelican, which must have been bred here, were found in the fen. The marsh-harrier, the bittern, the great bustard, the ruff and his mate the reeve, the black-tailed godwit, the

black tern, and the avocet have all disappeared from the
district, as has Savi's warbler, once common in the neighbourhood of Milton. Its curious nest, woven out of the
broad leaves of a reed, can be seen in the University
Museum of Zoology. It is well worth while to visit
this collection, where we shall see the species which used
to live here and learn to recognise the many beautiful
birds still left in the county. But to know the live birds
we must spend days in the fens, quietly hidden away, and
then we shall hear all the strange whistles and calls and
pipings of ducks, snipe, and redshank, the chick-chick
and chatter of the reed and sedge warblers mixed with
the rustling of the reeds and the long chur-rr-rr of the
grasshopper-warbler, called by fenmen the "reeler." Many
of the less common small birds can be enticed to nest in
gardens if boxes with the right-sized holes are provided.
The Norfolk plover still sometimes nests on the higher
ground, and on the river the kingfisher can be seen
flashing by. The red-backed shrike or butcher-bird is
often seen on the Roman road and Fleam Dyke.

Among reptiles the viper is almost unknown now in
Cambridgeshire, but the beautiful common or grass snake
with his golden collar and tapering tail may often be seen
gliding into the thicket, or diving into the water along the
edge of the fenland. There are several kinds of lizard
here, but they are not often seen. Four species of newt
are found, and newt's eggs, folded in the leaf of a water
plant, may be found in most ponds. The common toad
and frog occur everywhere, and the croak of the edible
frog, locally called the Dutch nightingale, and loud chirp

of the natterjack may be heard in certain localities.   The
edible frog has most likely been introduced.

The Swallow-tail Butterfly with its Caterpillar,
Chrysalis and Food

Cambridgeshire is rich in land and fresh-water shells.
One hundred and two different sorts have already been

found here. The brackish-water Paludestrina and the cockle live in the Nene and much remains to be done in searching the tidal rivers and central fenlands. We have already noticed the extinct shells of the gravels. Some others are dying out, like *Pomatias elegans*—a whitish spiral shell with a door which fills up the opening when the creature is drawn in. Numbers of its empty shells are found on the dry chalk land. The large snail eaten in France, called the Roman snail, though it is never found here with Roman remains, lives now on one small hill and nowhere else in the county. We have tried to reintroduce it in several suitable places, but without success. *Clausilia biplicata*, an exquisitely sculptured spiral shell, has recently been discovered alive near Cambridge in one place only. The smaller kinds of crustacea are plentiful in ponds and ditches, and the beautiful little crayfish is found in many of our streams.

Sixty out of the sixty-six British butterflies have been recorded from the county, but many have become very rare. The extinct large copper, last caught at Bottisham Fen in 1851, is recorded as a Cambridgeshire species (Tutt). Until Whittlesea Mere was drained, it occurred in great abundance there. The foreign variety has recently been introduced at Wicken Fen. The swallow-tail is still to be seen at Wicken Fen, though it is becoming more and more scarce, not so much because it is hunted by collectors as because the sedge to which the chrysalis is attached is being cut away everywhere, even down to the water's edge. Some years ago the writers rescued some hundreds in the chrysalis stage which the sedge-cutters had picked off the cut stems. They would have been despatched to dealers in

London for sale. We sent them to various large owners of fenland all over East Anglia to try and reintroduce them in suitable localities. They received every care and had their proper food, the marsh milk parsley (*Peucedanum palustre*) ; but in no single case did they reappear after the first year. The illustration, showing the life history of the swallow-tail, is drawn by one of our keenest and most skilful naturalists whose beautiful photographs teach more about nature than volumes of description.

We have nine hundred and two plants in our county out of the one thousand, nine hundred and six given in Bentham's *Handbook*, and ninety of these, including water or fen plants, wood plants, and dry chalk-land plants, are characteristic of the eastern counties which border the German ocean. We find that many of these plants are especially suited to the dry, sunny climate. Many are hairy, some have very long tap roots, and some a small leaf-surface, which means few openings by which the water can escape from the plant.

Cultivation has brought many dry-land plants into the fen district ; the true fen plants are now confined to Wicken sedge-fen and the two or three other small undrained patches by the river—at Quy, Fulbourn and Dernford, and Chippenham Fen.

The sedge, *Cladium mariscus*, is the main, though uncultivated crop of the fen ; it makes a beautiful durable thatch ; its sharp serrated edges keep birds from building and rats from burrowing into roofs made of it, and it is still sometimes burnt. The "sedge litter" is sedge and grass, used for bedding cattle.

Along the watery edge of the fen at Wicken the tall marsh fern *Lastraea thelypteris* still grows with the marsh orchis, pennywort, and marsh pea, and wherever the ditches come they are bordered with silver tufted reeds

Chopping Fen Litter, Reach

and here and there pink heads of flowering rush. In other fens it is still possible to find the grass of Parnassus, round-leaved sundew, butterwort, bog bean and marsh epipactis, stratiotes (water soldier), bladderwort and the limnanthemum, and in one or two places blue and pink columbine and larkspur. The sandy land on the Suffolk borders has an interesting group of plants. The white saxifrage, *S. granulata*, covers the ground in places. The dry chalk of the dykes is clothed with flowers from spring to autumn. First comes the pasque flower—*Anemone pulsatilla*. At Easter time and all through the summer the turf is sweet with thyme and gay with yellow rock rose, blue flax, milkwort, pink-budded dropwort, sainfoin, kidney vetch and viper's bugloss, and here and there a bee orchis, with a dancing accompaniment of butterflies overhead, graylings, skippers, chalk hill and Bedford blues, and a host beside. The grape hyacinth still grows in one or two places. The cowslips, known as paigles here, are extraordinarily abundant in the meadows, but the primrose is almost unknown in the north of the county and rare in other parts, where the woods are full of the true oxlip, which is one of our most interesting flowers. It has been shown by Mr Miller Christie that this *Primula elatior*, the "true" or "Bardfield" oxlip, probably grows only on a definite area in the woods of west and south-east Cambridgeshire, with the adjoining parts of Essex, Suffolk, and Hertfordshire, always on the boulder clay plateau and always separate from, though it hybridizes freely with, the primrose, and rarely with the cowslip. We have six different forms of primula in the county, three species and three hybrids.

The tidal banks of the Nene and Ouse give conditions which suit estuarine plants such as *Aster tripolium* and *Plantago maritima* and *Artemisia gallica*, var. *maritima*. A line of seaweed with dead crabs and small shrimps shows the high water mark. Some of these plants, such as *Scirpus maritimus*, still live at Littleport, and are thought to be survivors from the time when the tide flowed farther up the Ouse towards Cambridge.

Some of the local flower-names are interesting. Milfoil and shepherd's purse are sometimes called "break a mother's heart." *Hypericum calycinum* is "bird's nest," the long stamens forming a nest round the pistil. Bladder campion is "pudden bags." Willow herb "apple pie." Scabious "needle and pin cushion." Such names lead us to the study of charms and plant lore and the practical use of herbs for cures. The opium poppy was grown in the fens as a specific for ague and much evil resulted from its use. But in many a village there was some clever kindly woman who was at the beck and call of all in distress. She used both wild and garden plants for her cures; made salve of ground ivy and plasters of white lily petals; her garden was full of herbs and her cupboard of physic and "rubbing stuffs" which she administered with sympathy and confidence, free of charge, to the great comfort of the whole community. The memory of such a one is still green in Quy.

9. Climate.

Weather is the condition of the air at any time. It may be hot or cold, wet or dry, clear or cloudy, calm or stormy. Climate is the average of such conditions observed for a long time in any place. So we say, it is quite exceptional to have such weather in our climate.

To foretell the weather therefore, or to explain the climate, we must look to the same causes, for both are practically determined by the temperature and moisture of the air, which in turn are dependent upon the prevalent winds. The prevailing winds are due to the flow of air from a region of higher towards a region of lower pressure and these in our country, travelling north from equatorial regions and bent to the east, owing to the rotation of the earth, are the south-west winds, which carry with them the higher temperature and the greater moisture of the regions over which they have passed.

These winds also blow the warm surface-waters of the ocean with them and help to produce the upper currents of the ocean which have such a direct influence upon our climate. The one which affects us most is known as the Gulf Stream.

Our south-west winds therefore, which blow over the ocean from warm equatorial regions, are very moist, but the north-east wind, as it blows over land from cold northern regions, is dry and parching.

The atmospheric pressure increases from the south to the north of Europe, and this difference is greatest in the winter months; therefore the south-west winds prevail most in winter, carrying a high temperature and moisture with them, so that the winter temperature of the British Isles is much higher than what is due to mere latitude. For instance we should expect from its latitude that the mean winter temperature in Shetland would be 3° and that of Cambridge 17°, whereas in consequence of the warm currents of air and ocean, the mean winter tem-

perature of Shetland is 39° and that of Cambridge 37°.
The mean summer temperature of Cambridge is 48·9°.
This difference increases so much as we go north, that if
we draw lines on a map through the places where the
temperature is exactly the same we shall find that in the
British Isles these lines, called isothermal lines, run north
and south, so that we get to a colder and drier climate
not by travelling north and farther from the equator but
by travelling east, and therefore that the eastern counties
of England have the lowest winter average. When these
warm damp winds have reached the land their effect
depends upon the height and direction of the mountains
they have to cross. If the air has to rise into higher
regions of the atmosphere, where the cold and expansion
are greater, the watery vapour is condensed into cloud and
eventually falls as rain. Therefore a range running across
the path of the prevalent winds must drain them of their
moisture and produce a great difference in respect of the
amount of cloud and sunshine, drought or rain, on either
side of the mountains. The south-west winds have been
tapped by mountain ranges and high ground long before
they reach Cambridgeshire and the other eastern counties,
and therefore our average rainfall is small. The effect of
rain is very different according as it falls heavily for short
intervals or continues as a drizzle over long periods. It
has been known to rain a part of every day from the
31st of August to the 1st of December in a place in South
Wales where the average annual rainfall is not half that
at Keswick.

A great deal of the rain recorded in the east of

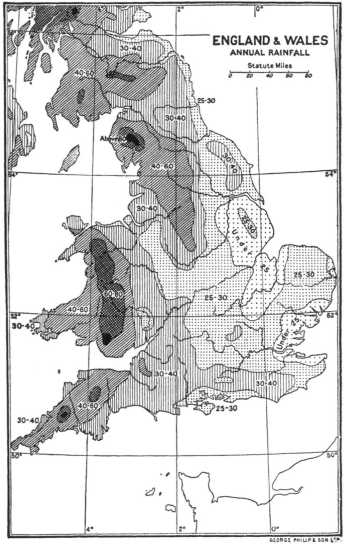

ENGLAND & WALES
ANNUAL RAINFALL

Statute Miles
0   20   40   60   80

30-40
40-60
25-30
30-40
Above 80
40-60
30-40
Under 30
25-30
60-80
40-60
30-40
25-30
Under 25
30-40
40-60
30-40
30-40
25-30

(*The figures show the annual rainfall in inches.*)

GEORGE PHILIP & SON LT.ᴰ

England falls in thunderstorms. In London, where the conditions are very like those at Cambridge, 3·12 inches of rain fell in two hours and seventeen minutes on August 1, 1846, and on April 13, 1878, a fall of 4·60 inches was recorded, whereas the average annual rainfall is 26 inches. The average rainfall at Cambridge is 22·81 and decreases to the north and east.

Woodlands affect the climate directly by retaining the damp and by shading the ground from the sun's heat and thus retarding evaporation. The tangle of roots also protects the soil from being washed down the slopes, the hillsides from being stripped of their soil and the river courses being choked with mud banks, impeding the flow and converting a valley with rich alluvial pastures into swampy wastes periodically covered with mud and gravel. Wide tracts of marsh land give off watery vapour which intensifies the effect of frost and affects the climate in many ways. The importance of understanding these agents is obvious when we look back at Chapter 7 *b*, in which we had to describe great terraces and spreads of gravel, the formation of which seemed to require rushes of water which neither the present slope of the ground nor the stream we see there to-day is sufficient to explain.

10. People—Race, Dialect, Settlements, Population.

The earliest race of which we have any traces left in Cambridgeshire are the people of the Palaeolithic or Old Stone Age who were here with the extinct animals whose remains are found in the gravels (see p. 64). We know nothing of them except from the rough-dressed, that is unpolished, stone implements they have left scattered in considerable numbers here and there on the surface or in the gravel, but these implements are like those found in other parts of the world, where a little more has been made out about the people who used them.

Next came Neolithic, or New Stone people, who had found out how to grind and polish flint and other hard stones. These seem to have come from the Mediterranean area along Western Europe, and some like to name them from the Iberi, who lived in Spain, along the route they seem to have followed.

Then we have mention of Gwyddels or Goedels, and Gaels and Cymry and Brythons, and these have left traces of their sojourn here and there in Britain, and some of them may have once lived in this district.

In the fens there was a tribe known as the Girvii, who were recognised as a distinct race even down to Norman times; and another tribe, the Iceni, occupied all the northern part of East Anglia and were such a powerful race that the Romans called them Cenimagni (Iceni magni).

6—2

We know that wave after wave of invasion broke upon our shore from the earliest ages. The North Sea rovers of Teutonic race—some Norse, or Scandinavian, some German—poured in as soon as the Romans had withdrawn. Then came Angles and Frisians and Saxons with their long sword-like knife called a seax, from which they take their name, and which we dig up sometimes in Cambridgeshire. All these races went to the making of the English people. We have already explained the division of the country into separate kingdoms, and how the line between East Anglia and Mercia ran through Cambridge.

Through some centuries we hear more of the conflict between the Scandinavian and German elements. But though we know that Danes occupied all this district, it is curious how few traces they have left of their language either in place-names or folk-speech. We have no places ending in "by" and "thorp" and "thwait," of which we have such an abundance in the north of England.

Then came the Normans with their Romance speech, and modified the Old English names and words.

If we try to make out the origin of the inhabitants of Cambridgeshire from the records of invasions, immigrations, or the introduction of large bodies of men from abroad to carry on great commercial enterprises, we shall learn that the population is made up of many different peoples, and if we wander about among the country folk, or the inhabitants of our towns and villages, and visit schools and places of concourse, we shall find that this impression is confirmed. We do not see any uniform type as the outcome of the mixture of many different races, but a great variety of types not blended, but grouped to a certain extent topographically.

In the Fens. Coming back from Work

What we do know is that there were dark-haired, dark-eyed and dark-complexioned people in this country who were mostly driven to the mountains of the north and west, though some small settlements of them remained sporadically in their old haunts; that the same dark race is still seen in France; and that Bretons came here to work and Huguenots took refuge here when driven from their own homes by persecution. We know also that the grey Scandinavian, the florid German, the flaxen Fleming, the yellow Finlander, and perhaps the red Votiak travelling along the Baltic, have all contributed to the population of the British Isles. Now when we see in Cambridgeshire a dark man who would not be looked upon as a stranger if he turned up on market day in any inland town in Wales, we may infer that he may be one of the pre-Roman Britons or a Breton who followed William of Normandy, or came in later times from the north-west of France to work on the drainage of the fens, but we know that he is not one of the North Sea or Baltic races.

Another common Fenland type is a sandy-red man, such as may be found in any seaport in Wales and more or less commonly all round the coast of England, Scotland, or Ireland. He cannot for a moment be taken for a man of Snowdon or a Cumberland man, nor is he quite like a lowland Scot or a Yorkshire man. He belongs to the Baltic type, and his ancestors may have come over with any of the earlier invasions of even pre-Roman times, or they may have been Norsemen who followed the Norman conqueror; for we must remember that the followers of William were not all dark, although in the Bayeux tapestry the English are coloured red and the Normans black, for distinction. William himself is represented in the picture at St Etienne in Caen, as a red man, and his son received the distinctive title Rufus. In later times large numbers of labourers were brought in from the Low Countries to carry out the embankments and drainage of the fens—

Old Style Skating Champions.
Turkey Smart and William See

work of which they had great experience in their own country—and in still later times the phosphate diggings and cement works drew many skilled labourers into the country from far and near. Thus we have plenty of historical evidence that people from many different countries have settled from time to time in Cambridgeshire, and the laws of Mendelism may some day explain the results we see.

All these different people now speak the same language with little peculiarities of pronunciation, especially in some of the vowel sounds, such as we commonly associate with the cockney speech. There are, however, few words in use which we can infer to have come down directly from pre-Roman Celtic. The talk of the people is in a language derived from various invaders, later than the Romans, modified now by education.

If we turn to the names of places it is interesting to note, though we must not push this too far, that most of the " hams " are on the south-east of the county, though many exist elsewhere. Most of the "tons" are in the south and west, spreading into Bedfordshire, Huntingdonshire, and Northamptonshire, that is to say through Mercia. London, being north of the Thames, its dialect partakes of the northern and eastern type, and is closely akin to that of Cambridge.

In some of the more important villages, not dependent upon and in constant communication with the larger centres, we still find many of the old words belonging to the folk-speech of earlier times.

The following local or unusual words—for most of which we are indebted to friends at Cottenham—are employed in Cambridgeshire, though it is not intended to imply that they are peculiar to the county :—

Scrawm = to scramble or scream (Norse); Scrawm thrush = the mistle thrush; Slub = thin mud (Norse); Fleet = shallow (O.E.); Flitt'n dish = skimming dish used for milk; Frem, used of lush vegetation (A.S. fremman, to advance, and Norse fram, forward); Dag = dew (Norse); Bawm = to smear; Bawm-slare = flattery; Beaver = a slight meal between breakfast and lunch (Norman); Bewst = "you can't touch me, I'm cried bewst" (Norman boiste); Ew = a mess (Norman); Mouze = "I wouldn't be mouzed up in the house this lovely day" (Old French); Turkis = to make an old garment smart again (Old French); Ash paddled = pale; Chimble = to cut or tear into pieces; Disagruntle = to disturb; Gotch = a pitcher; Horkey = harvest home and cf. Herrick, Hock-carts; Jill = a swing: to swing up and down; Paigle = cowslip; Quacken = to nauseate; Shooligay = lacking in energy; Stunt = blunt of manner; Stunt = very steep; Spalt = easily broken (Low German); Toible = a double-headed pickaxe; Twinet = a gimlet.

The name of St Etheldreda was shortened into Aldreda, and then Normanised into Audry or Awdry, one of our present Christian names. It has also given us the name tawdry, from the cheap ribbons called St Awdry laces, or, more briefly, 'tawdry laces' sold at the annual fair at Ely. Shakespeare and Spenser mention the latter and Drayton calls them "taudries, a kind of necklace worn by country wenches."

The population of Cambridgeshire was at the beginning of this decade 185,594.

11. Agriculture.

Agriculture has been carried on in this county from prehistoric times. The rude millstone tells of grain to be ground and the abundant remains of domestic animals suggest enclosures, even if we did not so commonly find ditches and banks associated with early settlements.

The Romans were skilled agriculturists, and the Saxons and early English generally had their methods of assigning portions of land around their villages for systematic cultivation.

Still there was but little land enclosed before the eighteenth century. The first "Inclosure Act" was passed just 200 years ago, and within the last century one could ride from Cambridge to any of the surrounding villages without seeing a fence. The effect of this is seen in the straightness of our roads. From the middle of the eighteenth to the middle of the nineteenth century was the time when almost all the cultivated land in England was enclosed and assigned to different owners.

From the return made last year by the Board of Agriculture we learn that the total area of land in Great Britain is 56,199,980 acres. Of this 26 per cent. is arable or ploughed land, 54 per cent. pasture, and 5 per cent. woodland. There remains about 8,000,000 acres which are unsuited for cultivation, or appropriated for houses, roads, etc.

Cambridgeshire is one of the smaller counties, ranking twenty-eighth in point of size and having an area of 551,466 acres, and we may approximately assign to each of the above industries a proportionate part of this area. Counties however differ much in position, climate, soil, and in the kind of crops which it pays best to grow.

For example, in the mountainous part of the country stock can be fed on the high ground, or moors, during the summer and brought down to the valleys in winter. Thus hilly districts, even when largely covered by heather and turf, are for practical farming purposes to be regarded as permanent pasture. On many such farms the supply of lowland pasture, or dry fodder, or roots, is insufficient to keep through the winter all the stock that can be grazed on the uplands in the summer, and some of it must be sold in the autumn.

On low-lying country farms, on the other hand, where a succession of crops is obtained, rye grass takes its turn, while winter food is provided by roots either eaten off the ground or stacked, cut up and measured out, or by straw and hay, much of it from sown grass and clover, in rotation. Even the ancient division into parishes is said to have been influenced by the necessity for assigning a due proportion of high pasture and low tillage, as is more clearly seen in Kent.

Cambridgeshire belongs to the latter group, for out of its 551,466 acres only 1547 can be called mountain or heath land used for grazing, yet the number of cattle was last year 58,920, and of sheep 173,823. The county being celebrated for the quality of the sheep originally bred by Mr Jonas Webb at Babraham. The area under clover, sainfoin, and sown grasses reserved for hay was 36,384, and not for hay 13,152. It can hardly be reckoned as a great horse-breeding county, though many cart-horses are bred in it, and the great racing centre of Newmarket encourages the breeding and training of race-horses. The last returns show that there were 33,700 horses, but of these 22,871 were used for agricultural purposes. Pigs are universally kept, the total return being 62,029, but there are no great piggeries.

The crops grown are chiefly for home consumption, and no large surplus produce is exported. A considerable quantity of home-grown wheat, however, is sent away, often in the very ships that have carried a still larger quantity of foreign wheat into our ports. But this is because the quality is different, and bread of the character which is chiefly in demand cannot be made from home-grown wheat alone.

There were 92,665 acres under wheat last year, the estimated yield of which was about 31·82 bushels per acre, and Burwell wheat is especially valued as seed. Barley (51,682 acres) and oats (51,734 acres) covered about the same extent of ground, but there were only 1161 acres of rye.

Cambridgeshire is one of the chief bean-growing counties, having about 21,064 acres under cultivation, while there is little more than one quarter of that area devoted to peas. Cambridgeshire is also the principal county for buckwheat, which occupies 1860 acres. It is chiefly used for feeding game, and the supply is so far short of the demand that in 1905, 140,860 hundred-weights were imported.

Potatoes covered 25,133 acres; turnips and swedes 13,946; mangolds 15,642; cabbage, kohl rabi and rape 6132; vetches or tares 3067; lucerne 2705; of flax there is very little, while hops are not grown at all in Cambridgeshire.

The climate of Cambridgeshire being one of extremes,

crops which have been retarded by the cold of winter, are
often quickly ripened by the summer's heat, so that the
average date of harvest is somewhat earlier than that
for the whole of England ; for instance, wheat harvest
commences July 28 and ends August 29 ; barley commences July 29 and ends about August 31 ; oats harvest
commences July 25 and ends about August 29; while
the earliest recorded dates for commencement are for
wheat July 14, barley July 17, and oats July 14.
Orchards cover some 4078 acres.

The recent foundation of a School of Agriculture
in the University of Cambridge will certainly bring about
changes which must soon produce an effect not only in
this county, but also throughout the whole of East Anglia.

There is now a Professor of Agriculture who not
only superintends the study of subjects bearing upon
agriculture, but also undertakes the conduct and superin-
tendence of agricultural experiments and inspections on
behalf of County or Borough Councils, or of other bodies
or persons.   There is a Professor of Agricultural Botany
whose researches into the evolution of useful varieties
of wheat have gained a world-wide recognition.   There
is also a strong staff conducting the work of students
in all subjects bearing upon the science and practice of
Agriculture, while the methods of detecting and com-
bating the various pests are receiving special attention by
skilled observers.   A Diploma, open to members of the
University and also to others, is granted as the result
of examination to those who have shown themselves
proficient in the subjects prescribed.

12. Forestry.

Cambridgeshire has only 6355 acres of woodland, the smallest area of any English county. Of this, 1196 acres consist of coppice (that is wood which is periodically cut down or severely thinned and reproduces itself naturally by shoots from the stools of the felled trees) and 449 acres of plantations (that is lands planted or replanted within the last 12 years or so) and 4710 of other woods. Now considering that £25,242,216 worth of timber was imported in 1905, it would seem worth considering what improvements might be introduced, not only by planting large tracts which cannot be more usefully employed in any other way, but also by a more careful selection and supervision of the timber sporadically occurring along our hedgerows, belts for shelter, or so-called waste places.

In Cambridgeshire all the circumstances are so diverse, that there is ample opportunity for observing the various growths and trying experiments, under the different conditions of soil, moisture, height, and temperature upon which the variations seem to depend. Cambridgeshire is thus admirably adapted for experiments in acclimatization and timber growing, and we may expect great results from the recent development of a strong school of Forestry in the University. There is already in the Botanic Garden a very instructive pinetum and a good collection of forest trees. The Huntingdon elm, a cross between *Ulmus montana* and *U. glabra*, is a characteristic tree of our district. Fine chestnuts are common.

Seldom can trees of larger growth be seen than those round Milton Hall, and at Pampisford Hall there is one of the finest collections of conifers in England. Beech often grows well on the low ground when the soil suits it, but it comes out naturally as the characteristic tree along the sides of valleys where, on the steeper slope, the chalk is exposed. You may even trace the chalk by the belt of beech trees which exactly follows it. On the loams below and on the clayey plateau above, where in old times the dense tangle of mixed forest and undergrowth formed an impenetrable thicket, the tendency is still to run into varied growth. On the wetter, springy spots are seen varieties of alder, of which in other districts clogs are made; birch, a useful wood for furniture; and willow, of which cricket bats, among other things, are manufactured. The pollard willows along the rivers furnish poles for various purposes, while the thin shoots springing from the stumps of shrubby species in the osier beds, furnish the material for basket-making. Where sand and gravel lie in irregular patches, conifers thrive; on the clay which is not water-logged oaks will grow.

There is no clear evidence as to when the great woodland of Cambridgeshire was cleared. We can hardly imagine that our county boundaries on the south-east could have been fixed while the forest still existed, for they could not have been traced or marked with any precision when the dense growth of wood extended continuously over all the plateau. Nor can we believe that the woods on the Cambridgeshire plateau were not cut down till the time of James I, when the continuation of our East Anglian heights, known as Bernwood or Brentwood, was cleared of timber. The work probably commenced very far back, and went on gradually down to comparatively recent times.

13. Special Cultivations.

A few crops which require special conditions for their growth are peculiar to certain districts and give a character to the whole country, as do the hop gardens to Kent and the orchards to the cider-producing districts.

Cambridgeshire used to produce several special crops, and some of these, which were cultivated in the earliest times, still survive. Teasles for dressing woollen goods are now partly superseded by metal carding machinery and in Cambridgeshire only grow wild. Mustard is raised in considerable quantities, especially in the neighbourhood of Wisbech.

The purple saffron crocus, which was used in early times by the Greeks and Romans as a dye and a perfume, gave its name to Saffron Walden in Essex, where it was cultivated for use in medicine and cookery. It was grown also in Cambridgeshire near Cherryhinton, and in the Isle of Ely, where there are fields near old farm houses still called Saffron Close. In 1763 an old writer says, "The fields near Cambridge furnish the town with the best saffron in Europe, which sells annually from 24 to 30 shillings a Pound." The name is all that remains— not even a stray crocus flower is now to be found in the fields.

But woad, which is always connected in our minds with the dyed and tatooed bodies of the Early Britons, is still grown in the fenland for dyeing purposes. The old woad mill, built of turf blocks arranged in the ancient herring-bone pattern, with a timber and reed-thatched roof, can still be seen at the village of Parson's Drove, six

The Woad Mill, Parson's Drove, Wisbech

miles from Wisbech. The plant, *Isatis tinctoria*, grows about six feet high, and has a blue-green leaf and bright yellow flower : the people still call it by its old Anglo-Saxon name, *wād*. Miss Peckover has written an interesting account of the industry. The young plants are delicate and the crop requires much care. It is weeded

by men and women clad in hardened skirts and leathern
knee-caps, who creep along the ground and take out the
weeds with a curious little hand-spade which fits into the
palm. The plant is picked by hand. The leaves are
crushed to a pulp in the mill by rude conical crushing
wheels dragged round by horses, and are then worked by
hand into large balls and laid on "fleaks" of twined
hazel, or on planks in special sheds for three months to
dry. After this the balls are thrown together, mixed
with water and allowed to ferment in a dark house for
five or six weeks. The woad is then rammed into casks
and is ready to be sold to cloth manufacturers.

Another industry which carries us back very far is
that of basket-making. This is one of our most ancient
crafts; the very name is Celtic, and is still used in
Welsh; while the Roman poet, Martial, mentions the
British basket (*bascauda*) in one of his epigrams:

"Barbara de pictis veni *bascauda* Britannis."

The wattle of the early huts was a kind of basket-
work, and the ornament on some rude urns suggests that
they were carried in basket-work, the pattern of which
was sometimes reproduced in the pottery itself. The
withies were cut from willows growing on the swampy
fens and river banks. Now small areas which can be
easily irrigated or flooded have been enclosed and in these
places, which are called osier beds, willows are planted
and carefully tended so as to produce long shoots which
are regularly cut, peeled and seasoned, and afford employment to large numbers of people for the manufacture of

Woad Cultivation near Wisbech

baskets and chairs, for which sedge and rush also are largely
used. Near Cambridge, at Ely, Over, Somersham and elsewhere, the osier bed and the basket-makers may still be seen.

Reeds and sedge, though not a cultivated crop, have
been used ever since the first rude huts were built. They
have to be protected from fire and flood, and cut and
gathered for litter and thatch at the right season. Their
care, preparation, and transport form an important industry,
giving employment to a large population in the fenland.
Bakers formerly used sedge to heat their brick ovens : in
the eighteenth century a hundred sheaves could be bought
for four shillings.

Another ancient industry was the capture of ducks
and wildfowl in the "decoys" where willows and
screens made of reed played an important part. A most
skilful and intricate trap was arranged in this way—tame
ducks were kept on the pool to entice the wild birds
down the curved approaches or "pipes" leading from the
pool for some fifty yards into the reed bed, and a well-
trained little dog, "the piper," encouraged by a bit of
cheese, jumped, or ran round one low barrier or "leap"
after another, to make the wild duck go farther and farther
up the pipe, which was covered in by a net supported on
poles. The men who watched them were hidden by the
screens, and gradually the birds were decoyed to the small
end of the pipe, where there was a bag net, which was
suddenly twisted and the birds were caught.

Carter writes in 1753 : "In these fens are several of
those admirable contrivances called decoys, in which it is
incredible what quantities of duck, teal, wigeon, and all

kinds of wild-fowl, are taken every week during the season. There is one near Ely which lets for £500 a year, and from that alone they generally send up to London 3000 couple a week."

14. Industries and Manufactures.

Of the industries connected with farming, cheese-making was at one time a speciality of Cambridgeshire and Cottenham cheeses were famous, but this industry has diminished of late years.

Fruit-growing, on the other hand, is an increasing industry and its success is specially marked at Histon where it was introduced about a hundred years ago. The business, which started in a simple way, was so successful that it soon required new premises, and it is now carried on in a large factory fitted with the best modern machinery, where over a thousand workers are employed all the year round, while three thousand acres are under cultivation. The quality and quantity of fruit grown in the district have been greatly improved in this way and employment is given to village people. Fruit-canning in this country first began here, and has developed into the largest and most successful industry of its kind in England. The firm does everything for itself; it makes its own carts and its non-corrosive tins, it does its own building and lighting. Bees were kept at first as a means of fertilising the blossoms, but now the honey produced by them amounts to many tons annually, and

Gathering the Strawberry Crop, Histon

some of the villagers employed in the factory make a profit of from £20 to £40 a year by their bees. At Wisbech cheap trains bring hundreds of the poor people from the London slums to gather the crops of strawberries, raspberries, and gooseberries, but the workers employed at the Histon factory all come from Cambridge and the surrounding districts.

Land in various parts of the county has been specially prepared not only for fruit-growing, but also for asparagus culture, and supplies of this vegetable are sent to the large markets.

Formerly nearly every village had its windmill or water-mill for grinding corn, and many of these are still used for this purpose, though the import of flour already ground makes them no longer a necessity. Still, a large quantity of corn is ground and oilseed crushed in the county.

Rough and glazed pottery was made at Ely in the sixteenth and seventeenth centuries; and one pottery (Robert Sibley's) was not closed until 1864.

An interesting manufacture is that of shamoy (chamois) leather and parchment, which has been carried on at Sawston for over a hundred years. Like that of paper it was no doubt started here because of the good supply of water and easy means of transport.

The Sawston people in this factory are all skilled workmen and some of them have been employed here for forty or fifty years. Each man must be an expert in his own department because here everything depends upon the judgment of the individual, and any one of the fifteen or sixteen processes through which the skins go before they are turned into velvety "shammy" leather, can be made useless by one careless act. Each man's mind is bent on keeping the skins "in condition."

Parchment Making, Sawston

The old terms used tell that this is an ancient industry. The *pelts* come from the *felmonger* and are laid in lime and water, a *flesher* scrapes the skin with a special curved knife as it hangs over a *beam*—a block made of a tree trunk cut in half lengthways. The skin is *couched*, all wrinkles taken out. Then a clever machine splits it, dividing the wool or *grain* side from the mutton or lining side. The lining is *frized* by hand to remove the inner layer of fat and is limed again, and, when in condition all the lime is washed out. It is then squeezed in a hydraulic press, punched by heavy machinery in the *stocks*, until all "the water is killed," and treated with cod liver oil. After being dried and laid in wooden tubs, fermentation takes place and workmen "turn the heats," for if the skins lie too long here, they burn. An alkaline wash cleanses the skins of fat, and after they are dried some of them are bleached with sulphur, but wise housekeepers buy the brown unbleached skins which last longer and do not turn the silver black. In the *grounding* shop the leather is pared with a moon-shaped knife and smoothed with a wooden tool, called a *scurfer*. Hanging in this shop are small bunches of butcher's broom. The grounders gather it at Whittlesford and call it by its old name, "knee-um" (knee-holm or knee-holly). They find that its sharp leaf-branches sprinkle water on to the leather better than anything else. In the last process the beautiful soft skins are rounded and trimmed, then put up into bundles of 30, called a *kip*, and sent off to all parts of the world.

For parchment the *linings* are tied in a frame by strings fastened round grooved pegs, on the same principle as a Spanish windlass. The little balls made of scraps of parchment, which give the strings a firm grip on the skins, are used over and over again and are in this way covered by many layers. They are called *pippins* and are quite hard.   After being scraped with a *half-round* knife, dried, shaved, dabbed with whitewash and heated in a stove to remove the grease, the skins are then scalded and rubbed smooth with pumice until they are fine and smooth and ready to be dried and cut into sheets and sent out to law-stationers, book-binders, etc.   The parchment workers wear clogs, sheep-skin leggings, and *basil* aprons. A *basil* is an unsplit, tanned sheep-skin.   In this well-managed factory all the refuse goes to make soap, glue, dubbin, or manure for fruit trees and not one scrap of material is wasted.

The paper-mill at Sawston is the twenty-fifth built in England.   It has been at work for quite a hundred years and is in private hands.   The work people, 150 men and about as many women, are housed in the neighbouring villages and are particularly well cared for.   No brown paper, only paper for ledgers and fine writing-paper, is produced here, and the out-put is 20 tons a week.   One of the paper-making machines is the very earliest intro-duced into England, but others employed here are of the most modern type, and this is one of the few mills which still finishes the paper by copper glazing—a process which entails hand work and produces the best result.   The surroundings of the mill, the river, meadows, and fine trees are very charming.

Book-binding has for long been an important industry in Cambridge, and is still carried on by some excellent firms and highly skilled private binders.   In 1794 John Bowtell was esteemed the best book-binder in Cambridge. Specimens of his work may be seen in the University

Library, and some of his bindings have found their way into the Royal Library at Copenhagen.

Printing is an ancient industry for which Cambridge is famous. More than 300 people are employed at the University Press, and a considerable amount of printing is also done in many towns in the county. The right of

The University Press

appointing three stationers or printers was granted to the University by King Henry the Eighth in 1534. John Siberch, the friend of Erasmus and the earliest of these University printers, had set up a press in 1521 in a house bearing the sign of the King's Arms, where part

of Gonville and Caius College now stands.   His press
has disappeared, but nine of his books have come down to us
and some of his initial letters were used by the Press quite
recently.   Thomas Buck was another famous Cambridge
printer who worked in the days of Charles the First.

The religious controversies of the sixteenth and seventeenth centuries interfered with the work of the Univer-

Part of Machine Room, University Press

sity Press so that the printers had no peace or freedom
until the Commonwealth Parliament recognised the
University as a privileged printing place.   But it was
Richard Bentley, the great scholar and critic, who con-
ceived the idea of a University Press.   He appealed for
money to build a new printing house, bought beautiful
new type from Holland, and with a competent printer,

under the direction of the University, the new Press was established. This was in principle the Cambridge Press of to-day, where books held by the University itself to be of value are produced in a way worthy of them.

In the fifteenth century an old inn, " The Cardinal's Cap," occupied the site of part of the present University Press. The land was acquired and the buildings were begun at different dates ; some in 1804 and others in 1824. The quadrangle was completed by the erection of the present frontage and tower, built out of surplus funds subscribed to commemorate William Pitt, and the building is sometimes called after that great statesman. The Cambridge Press prints in every European language and many Oriental and other languages besides. Its productions are known all over the world, and its work to-day worthily maintains the great traditions of the past.

15. Mines and Minerals.

There are no mines in Cambridgeshire nor any minerals of such a kind as is usually understood when the word is used in connection with mining, but, if the meaning of the word is extended, there are several important industries dependent upon the digging and preparing of minerals.

Cement is a term applied to various substances, but that which is manufactured so largely in Cambridgeshire is a mixture of lime and clay, which has the property of hardening rapidly even under water. It used to be made

chiefly in the Medway valley, in Kent, where the alluvial
mud of the river was mixed in definite proportions with
the chalk procured from the hills on either side. By
and bye, however, it was noticed that the marl, about
70 feet of which occurs at the base of the chalk, was
composed of lime and clay in just the same proportion,
and works were established to manufacture cement from
the chalk marl, which crops out all along the valley of the
Cam. This introduced a new industry. But the success
of the earlier ventures caused too many cement works
to be opened, so that the supply exceeded the demand.
Moreover, in some places difficulties arose from the marl
not being of exactly the same composition throughout.
The strong companies however survived, and the industry
provides much well-paid labour and has contributed largely
to the prosperity of the county. The Saxon and Norman
Cement Works, near Cambridge, now turn out over 2000
tons a week.

There are several bands of hard, jointed chalk in the
marl which are of great importance, as they are the water-
bearing horizons, when below the level of saturation. In
the pit at the Bottisham Lode Cement Works (see opposite
page) there are two such rocks, which yield a good deal
of water. In the pit at the Saxon Cement Works they
occur—the lower, which is from 3 to 4 feet thick, at
about 12 feet from the base, and the upper, which is from
$2\frac{1}{2}$ to 3 feet in thickness, at about 22 feet; but these are
dry, as all this part of the chalk marl is here above the
level of saturation. The workmen in the figure are
standing on the Cambridge greensand or phosphate bed.

Just about the time when a great advance had been
made in agriculture by the discovery that we could put
into the soil the chemical substances wanted for the
growth of plants, Professor Henslow pointed out that
round Cambridge there was a deposit which contained

Pit in Chalk Marl, Bottisham Cement Works

an enormous quantity of one of the most useful in-
gredients, namely phosphate of lime, and it was suggested
that this might give the same kind of superiority to
England in agriculture that her rich coalfields had given
her in industry and commerce.   The phosphate of lime

occurred in small lumps and fossils, not only in the gravelly beds of the Lower Greensand, but also in a thin bed resting upon the Gault and forming the base of the Chalk. Farmers were comparatively well off then and bought the new fertiliser freely, so that from about 1850 to 1880 the industry thrived and it paid speculators to dig these "coprolites," as they were called from a mistaken idea as to their origin, even when 12, 15, or as much as 25 feet of overlying marl had to be removed to get at the 8 to 14 inches in which they occurred, while they were willing to pay as much as £150 per acre for the right to dig them, stipulating that they would put back 18 inches of the surface soil and restore the land to a state suitable for cultivation.

The process of converting the phosphate of lime into the superphosphate, in which form it is used for agricultural purposes, is somewhat complicated, and requires a great deal of grinding and application of chemicals, all of which involves extensive premises and skilled labour.

But with agricultural depression and the introduction of other chemicals more easily procured, the demand for the phosphate nodules decreased, while the depth at which they occurred and consequent expense of obtaining them increased as the shallower beds got used up, so that soon the industry came to an end and a large body of men, accustomed to earn high wages, were thrown out of work.

As might have been expected from the scarcity of stone suitable for outside building and the abundance of clay, brick-making is a very important industry, especially in the neighbourhood of Peterborough and Whittlesea.

At Cambridge, bricks are made of the Gault which generally burns white; at Ely, they are made of Kimeridge Clay, which burns red; and at Peterborough and Whittlesea, of the Oxford Clay, which also yields a red brick.

Although, speaking generally, there is very little rock available for building purposes, there are here and there small bands of somewhat irregular occurrence which furnish a useful material for many local purposes. The limestones of the Jurassic rocks are not of much account, yet garden walls are built of the Elsworth Rock, for instance, and, as has been pointed out, the septarian nodules of the Kimeridge Clay are used for making up the river banks.

The hardened portions of the Lower Greensand furnish an easily-dressed stone, known as Carstone, which, if properly chosen and laid in the manner in which it best resists the weather, improves with time. It has been much used at Sandy and at Ely for houses and churches, but there is no great quantity of it and it has seldom been carried far. Chalk has been used as a building stone from very early times. The hard, somewhat clayey band known as the Burwell rock, which occurs near the top of the chalk marl, is the best adapted for this purpose, and therefore we find quarries all along its outcrop through the county. Near Reach there are very large quarries which have been worked since Roman times, for " clunch "—as chalk used for building is locally called—was used in the basement of a Roman villa near where the Mildenhall Railway crosses the Devil's Dyke. The richly carved tracery of the Lady Chapel at Ely, and the internal work of Burwell and Barrington churches, are examples of its use. It is suitable for external work also, where it is not exposed to dripping water or continued damp, but it lasts well even for outside work if it is protected from the weather by coats of whitewash. Chalk is also largely quarried for burning into lime for mortar or laying on the land as a fertiliser. In the upper part of the chalk, layers of flint occur, and these are of great commercial value. They are largely used for building purposes, either in the rough form in which they occur in the chalk, or neatly dressed into cubes, and built in various patterns, often alternating with brick or stone. The flint itself resists any weather action, but the pieces are small, and, unless the mortar be so strong as to unite the whole into one solid mass, there is danger of its falling to pieces, seeing that the flints do not sit like bricks which might stand without mortar. Flint has been used from the earliest ages, of which we have any record, for making implements, and in the adjoining county, long after flint guns and flint and steel had gone out of use in this country, the flint-knappers found employment in the manufacture of gun-flints for China and Africa.

16. Fishing.

Though there is no sea coast in Cambridgeshire, and therefore no fisheries or fishing stations, as the terms are usually understood, still fishing is an important item among the productive industries of the county. From the earliest times, as we have seen, the fens were famous for their abundance of fish, and in the twelfth century we hear of the netting of "innumerable eels, large water-wolves, with pickerels, perches, roaches, turbots and lampreys, which are called water-snakes." "So great store there is here of fishes that strangers coming hither make a wonder at it, and the inhabitants laugh thereat to see them wonder."

The greater facilities of transport have now introduced the more abundant and delicately flavoured sea fish far inland, but Cambridgeshire is still a county of anglers. Many of the fish taken in our meres and rivers are still eaten, and even where the first object is not to obtain a supply of good food, the love of sport is a sufficient attraction to insure the protection and capture of large quantities of fresh-water fish. Trout are common in the upper rippling waters of the Cam, and have often been re-introduced where they have become scarce. Occasionally an exceptionally large individual has been taken in some of the deeper pools, as for instance one of 13 lbs. recorded from Newnham Pool, where one weighing 4½ lbs. was taken last year. Eels were plentiful according to all old writers, and the historian Bede's derivation of Ely as the isle of eels still finds favour with many. It must have been well known that eels were easily procured, otherwise the order given Dec. 10, 1220, to the bailiffs of Cambridge to supply 5000 eels by Christmas, for the use of the King while staying at Oxford, would have been extraordinarily harsh. They were probably taken in large quantities in eel-traps set in the path of their annual migration to the sea, and therefore certain localities were apt to be of exceptional value. This explains the protest of the Abbot of St Edmundsbury against the alteration of the course of the Nene, which was intended to protect Wisbech and the surrounding district from floods. The curious life-history of the eel has only recently been made out. The pike or jack is another fish much valued for sport and food. It occurs everywhere and runs to a great size. In former times, enormous quantities of pike were sold in Cambridge market. There is a habit of the pike which seems to suggest a case of instinct at fault. In flood they wander up small ditches and over meadows in search of food, but do not always take proper precautions about getting back, so they are frequently left in small isolated ponds, or even in holes dug on purpose to induce them to stay till too late, when the floods are subsiding. Many of the fish taken in the county are sea fish that have come up with the salt tidal water, which in old times was not held back so effectively by sluices and locks. But some, as the salmon and sturgeon, for instance, are fish which commonly run up suitable rivers. It is doubtful whether salmon ever occur now in any Cambridgeshire river, but sturgeon still find their way through, now and then. A large specimen was seen last year in the Ouse, near Over. Coarse fish are abundant, and takes are sometimes measured by the stone instead of by the pound. The quantity is however much reduced from what it must once have been. This may be explained by the drainage system now carried out. Much of the water from the great network of drains is now pumped up into the larger channels and runs out to sea, and thus a restriction is placed upon the free migration of fish in search of food and suitable conditions of life. There is very little destruction now caused by otters, or by birds and other animals which feed upon the spawn, as all of these, like the fish, are far less numerous than they used to be. The common fish of our rivers, besides the above-mentioned, are chiefly perch, carp, roach, dace, tench, bream, and some small fish like loach and bullhead used for bait ; and in the lower reaches, flat fish. Forty-two species, including the salt-water fish of the Nene, are found in the county.

17. Shipping and Trade.

Although Cambridgeshire has no seaport of its own, King's Lynn, situated at the mouth of all the navigable rivers in the county, may be regarded as the ancient port of Cambridge and of all the south and eastern parts of the county. So Wisbech, near the mouth of the Nene, was the port for all the north-western district. We get a good idea of the river traffic to Cambridge about a century and a half ago from the description given in an old book called *Cantabrigia Depicta*, published in 1763, in which we read: "The purest Wine they receive by Way of Lynn: Flesh, Fish, Wild-fowl, Poultry, Butter, Cheese and all Manner of Provisions, from the adjacent Country: Firing is cheap; Coals from Seven-pence to Nine-pence a Bushel; Turf or

rather Peat, four shillings a Thousand; Sedge with which
the Bakers heat their Ovens, four Shillings per hundred
Sheaves: These, together with Osiers, Reeds, and Rushes,
used in several Trades, are daily imported by the River
Grant.   Great Quantities of Oil, made of Flax-Seed,
Cole-Seed, Hemp and other Seeds, ground or pressed by

The Port, Wisbech

the numerous Mills in the Isle of Ely, are brought up
this River also; and the Cakes, after the Oil is pressed
out, afford the Farmer an excellent Manure to improve
his Grounds.   By the River also they receive 1500 or
2000 Firkins of Butter every Week, from *Norfolk* and the
Isle of *Ely*, which is sent by Waggons to *London*."

The rivers of Cambridgeshire with their locks and

sluices, towing paths and artificial cuts, are in direct communication with the canal system of central England. The "levels," the primary object of which was to carry away the water lifted by pumps from land-drains and ditches, are also canals along which "lighters" and barges carry the local traffic to and fro. These various waterways were the chief trade routes not only for local traffic, but also for stone from the quarries of Northamptonshire or Rutland, or for coal brought by sea from the north of England, or for Portland stone, or Purbeck marble from the south coast. The rivers were crowded with all kinds of vessels, propelled by oars, or sails, or towed. In old times, before the incoming tide was checked by sluices and locks, and when the upland waters were only let off with the ebb, these vessels had to wait till the current allowed them to ascend or descend the rivers, and sometimes there was a great block of craft waiting for favourable conditions.

An old writer describes Wisbech as "on the utmost Northern Borders of the Isle of Ely, from which town it is distant about twenty miles among Fenns and Rivers": but he says "it is the best Trading-Town in the Isle, having the Conveniency of Water-Carriage to *London* ......it sends to *London* every Year 5000 Tuns of Oats, 1000 Tuns of Oil, and about 8000 Firkins of Butter." Wisbech is still an important seaport. Ships of 3000 tons can come up to it. According to the latest returns, the tonnage of steamers and sailing vessels entering the port was 23,752, of which 10 were British with a tonnage of 4001, and 39 foreign with a tonnage of 19,751, while the voyages made in the coasting trade were 162 with a tonnage of 13,472. Now, however, the sea-going vessels are met here by a branch line from the Great Eastern Railway to the harbour, by the Midland and Great Northern Joint Line from Peterborough, with a tramway into the Market Place, and a steam tramway from Wisbech to Upwell.

Nor is it only by the rivers that this great traffic has been carried on. On the artificial cut, known as the New Bedford River, or Hundred-foot River, there was in the last century so great traffic that at Mepal, for instance, as many as 40 lighters would be lying off the wharf. There was a brewery there which is now a ruin, but which at that time employed sixty workmen, and obtained all its supplies and sent off all its produce by water.

The Saxon word Hyth or Hithe was applied to places along the river where there were facilities for loading or unloading cargoes, and there was often combined with the name a description of the use to which the particular hithe was assigned, or the name of the person or body to whom it belonged. So that we find "the Hithe" at Chatteris, and Clayhithe, 5 miles north of Cambridge: at Ely we have Brodhithe, Castelhithe, Monkeshithe, Stokhithe; and in Cambridge, Corn Hithe, Flax Hithe, Salt Hithe or Salters Hithe, Garlic Hithe, and Dame Nichol's Hithe.

The changes brought about by the diversion of the water into new channels have also affected the trade of many a town in the fenland. In the interesting *Records of a Fen Parish*, we read that there is in March "a bridge

of a single span crossing the river Nene—river only by
courtesy, for it differs in no respect from an ordinary
canal, save that its winding course marks it out as one
of the ancient water-courses of the country.   In days
gone by this waterway was of considerable importance
to the little town.   All heavy traffic was conducted by

Hundred Foot River, looking N.E., Mepal

these silent highways, and many a family in the seventeenth and eighteenth centuries acquired prosperity and
even wealth by conveying wheat to Lynn, and returning
with their barges heavily laden with timber and 'Sea-cole.'
Dear enough, too, to the buyer, was the 'Sea-Cole.'   In
1706 the Churchwardens paid 27 shillings per chaldron
for it, to give away under some benefactor's will to the

poor of the town. But the old industry has vanished,
driven **away** by giant steam, and the 'Hive' as the branch
canal which ran a quarter of a mile up the High Street
was called, where the barges were unladen alongside of
their owners' houses and barns, has been bricked over and
turned ignominiously into a sewer."

As our roads and railways now determine the position
of new towns and villages, so in old times the facilities
offered by the waterways made people build their houses
by the hithes and at the ends of the lodes.   Swaffham
Lode has still the decayed remains of an ancient landing-
place, and a large part of the village known as "Commercial
End" was built close to it.   At the landward end of
Bottisham Lode a village known as Bottisham Lode, and
larger than the original Bottisham village, has sprung up.
At Burwell, a very ancient place, a cross lode is carried
at the foot of the higher ground to serve the various parts
of this very long village.

Reach, however, seems to have been the most important place in this part of the district, and sea-going
vessels of considerable tonnage came up to it.

It is not safe to infer from such a name as Littleport
that the place was anciently a shipping station, when we
have also such a name as Newport on high ground in the
southern part of the county and a possible derivation from
either *portus* or *porta*.   There is just the same doubt in
the north of England respecting the origin of names
ending in " gate " or "yet," which may come either from
gate = *porta* or from gait, a way or road. The rivers which
were in the past the chief—almost the only—means of

communication are still largely used for goods in respect of which time is no object, or for goods which are likely to be injured by the jarring of transport by road or rail. There is therefore still considerable use made of the waterways for the carriage of local produce and manufactures, such as bricks and tiles, of peat, timber, and sedge, of clunch and clay, and of all kinds of agricultural produce and requirements.

18. History.

During the Roman occupation of Britain, that part of the country which is now called Cambridgeshire was, as we have seen, inhabited by a tribe called the Iceni. Little can be said of them except that they were tolerably friendly to the Romans, until the propraetor Ostorius fought and crushed them in a great battle of the Dykes, which probably took place near one of those huge earthworks.

After this the Romans began to abuse their power, and Queen Boadicea led a great rebellion of all East Anglia in which the Iceni took part, destroyed the 9th Legion and sacked London. The Cambridge villages suffered greatly when the revolt was put down.

Discoveries of pottery, coins, etc., which have been made, seem to point to a time of peace after this outbreak, and to the existence of small traders and workers on the land. The coin of the usurper Allectus found in the Cam valley shows a warship of the first British fleet, cleared for action.

The Roman influence was good, and when it was withdrawn, the Britons in the beginning of the fifth century felt the loss. Pirates sailed up the rivers Ouse and Cam into the heart of the country. Christianity was driven out, and all the old towns were destroyed. No Cambridge town survived. In the sixth century the East Anglian kingdom was formed, whose king was a Christian, but Penda king of Mercia was a heathen, and routed Anna, the East Anglian king, at the Dykes. Anna, who died fighting for his faith and country, was

Coin of Allectus, found in the Cam Valley, showing a Warship cleared for Action

the father of a famous family. Their home was at Exning, a spot which commanded the dykes and was in touch with the fenland. One of his sons became Bishop of London, another king of East Anglia, and five of his daughters became abbesses—one being Etheldreda, who first married Tonbert, prince of the fenland Girvii, and afterwards Ecgfried, king of Northumbria. The two Christian kingdoms of East Anglia and Northumbria were thus united through her. Later, as we know, she fled from the North and forsook the world to become foundress of Ely. She was succeeded here as abbess by her sister Sexburga, widow of the king of Kent. The eight chief acts of St Etheldreda were carved by Alan de Walsingham in the beautiful corbels under the great octagon.

Continual wars went on between these countries of

Archbishop Wilfrid installing St Etheldreda
as Abbess of Ely

East Anglia and Mercia, and for a time Mercia was supreme over all England, south of the Humber. The first inrush of Norsemen took place in 787: Cambridge was a strong military position, and was used by these Norsemen as a centre from which to plunder the country. In 870 they crossed the river at Cambridge, trod down the crops, ravaged the country and destroyed Ely. St Edmund, king of East Anglia, was defeated and murdered by them. In 878, by the Peace of Wedmore, this part of the country was surrendered to the Danes, but in 905 King Edward, son of Alfred the Great, once again invaded it because the Danes were helping a revolt against him. A great battle was now fought at the Dykes, and the Danes were defeated. A few years later the Danes submitted to Edward, who organised Cambridge as a county for the first time. At the end of the tenth century, the Danish invasions began again. By the battle of Maldon, the East Anglians drove back the Danes. Meanwhile, the county was gaining a strong separate position for itself, apart from East Anglia. It was led by strong men, and took a leading part in the Danish wars. At the battle of Ringmere, in 1010, when the host of East Anglia fled from the Danes "there stood fast Grantabrygshire alone." The county suffered terribly in consequence. The town was sacked and burnt, and the people massacred. Canute beat the forces of England at Assandun, which may be Ashdon on the borders of Essex and Cambridgeshire. This battle decided the fate of England, and he was the first Danish king to rule over the whole country. His reign was prosperous, and the monks at Ely were encouraged by him. The old ballad gives a picture of his visit to Ely for the great Feast of the Purification :

> Merrily sung the monks within Ely,
> When Canute, the King, rowed thereby.
> "Row, my knights, row near the land,
> And hear we these monkës song."

But, on the whole, Cambridgeshire suffered under the Danes. The prominent part taken by it in the wars had lost for it many leading men. The country also had suffered. The land was often ravaged and burnt, and the crops destroyed. The battle of Ringmere had left its mark. All this is shown by the condition of the country at the Norman Conquest. Since Assandun, the leading families had disappeared. No thanes (leaders) are mentioned. The land was seized by William for his own subjects.

Yet Ely was still a strong military position. The Danes came down from Yorkshire in 1070, and forced the people to resist. The island of Ely became a camp of refuge to all those who had been deprived of their lands, or who were discontented. The movement took a strong hold on popular imagination. The difficulty of approaching Ely, owing to the morasses of the fens, the good food supply, all made it an ideal place for a last stand. Hereward the leader resisted all attempts here, but William himself finally entered Ely by a causeway made over the marshes south-west of Aldreth, where the dry land on the south approached most nearly to the island. He was able to do so largely owing to having threatened and bribed the monks to help him. This was the first time Ely was reached by a causeway. Things improved for Cambridgeshire with the Conquest. Churches were rebuilt, the Abbot of Ely was now made bishop, and the county settled down once more. The growing prosperity and trade are shown by the existence of a body of Jews in Cambridge, who always appeared at this time

where the country was rich. Under Stephen the lawless-
ness of the barons was felt everywhere. Geoffrey de
Mandeville, one of the worst of them, seized Cambridge
and Ely, and plundered the county. He was finally shot
by an arrow when attacking a castle built at Burwell by
Stephen on purpose to keep him in check.

In the reigns of John and Henry III, Cambridge once
more became the centre of the discontented, who formed
a party and revolted against the king. In 1215 the
town was sacked by royal mercenaries—the hired scum
of Europe—but the forces of the barons who were fighting
against John suddenly appeared, and captured the Norman
castle of Cambridge and the ford over the Cam, which at
that time was a point of great strategic value. John was
forced to retreat to the north of the county and over the
Wash, where he lost all his jewels and his crown. It
has been said that it was in crossing the Nene estuary
somewhere near Wisbech, that this happened. John was
defeated, but the conflict between the nobles and the king
went on, and in 1265 Ely was once again used as a last
resort of the revolutionary party. The Pope supported
the king, but the clergy at Ely were not willing to do the
same, unless he allowed the Church to be free from State
control. At last Edward I succeeded in capturing Ely
by crossing the fens on bridges of hurdles. He used
Cambridge as his base. During this period several important events had occurred in the county. The first
beginnings of the University were seen, though there
was little organisation. The chief characteristics of the
time were the strength of the nobles, who practically had

power of life and death in their districts, as is seen by the number of gallows found in various villages; also the great influence of religion at the time in the country. The monasteries were numerous, and it was there that the poor were educated. Many new churches were built,

The Old West River, where the Causeway crosses it

and the religious duties of the people were carefully exacted. The clergy had great power in the county. In the time of the Black Death, 1348–50, vast numbers of the population were swept away. Wat Tyler's rebellion followed, but was not dangerous in Cambridge. Several places were sacked and the charters of the colleges were burnt, but the Bishop of Ely was able to suppress the revolt. The Wars of the Roses hardly touched this county, which was in a very prosperous condition, with wages high and work plentiful. The University was extended, and King's Chapel begun.

All through the Reformation this county played an important and very consistent part. It was the centre of the new thought; new learning was taught and the idea of the University was to encourage new ideas, not—as in the monasteries—to enforce old ones. About this time pensioners were admitted to the colleges; they were students who paid fees: before this there were only scholars and fellows. The new ideas of the Reformation were seen when people began to endow the colleges, instead of the abbeys, with money. Then Henry the Eighth dissolved all the monasteries. It was at this time that the beautiful sculptures in the Lady Chapel at Ely were injured, and that the religious houses of Thorney, Chatteris, and Shingay were destroyed. The colleges used the remains of many of these religious houses to enlarge their own walls. Meetings of the Reformers took place at the White Horse Inn (now the Bull) in Cambridge, and Cranmer, Latimer, Ridley, and Cox of Ely—all leaders in the movement—were Cambridge men. The county suffered at this time owing to the uncertainty of everything, and owing to the abolition of the chantries where the poor had been taught. The churches were despoiled, and the parish accounts of March give an interesting picture of the confusion of the time, for under Edward VI the altars were destroyed; under Mary they

were set up again ; under Elizabeth they were again
destroyed. Under Mary there was a reaction towards
Catholicism ; yet she was surprised by her enemies at
Sawston Hall, and only escaped on a pillion, riding behind
one of the servants. She was proclaimed Queen in Cambridge market-place, where a few days earlier there had
been rejoicings over the proclamation of Lady Jane Grey.

Under the Stuarts and during the Civil War the
county was strongly Protestant. A riot took place at
Melbourn when Charles I tried to levy the unjust tax
of ship money. Cromwell fortified the town of Cambridge, and made the shire into one of the seven associated
Protestant counties. Twice the king marched on the
town, but both times retreated.

On 10th June, 1647, the Parliamentary army, discontented at not being paid, gathered at Triplow Heath
under Cromwell, and brought away the king from Holmby
House in Northamptonshire, where he was then im-
prisoned, into their own custody at Newmarket, purposing
to make use of him against the Parliament. Cromwell
then threatened to attack London and marched by Royston
to St Albans with the king as his prisoner. Two years
later Charles I was executed.

With the Restoration, the modern period began in
the county. Since then the fens have been thoroughly
drained. The southern part of the shire, once all open
field, has been enclosed ; regular coaches, and then railways, have taken the place of the former methods of
communication. The University has expanded and been
opened to all, and the education of the poor in the county

has been put under State control. Yet life at the present time is hard for the country people. Agriculture is bad; many of the new industries have failed, or do not pay well. Coprolite digging, which brought such wealth to the county, is over, and wages are low. The two small towns of Royston and Newmarket have gone over the border into the richer counties of Hertfordshire and Suffolk. The middle of the nineteenth century was a time of great prosperity, the end, one of depression.

19. Antiquities.

There are few counties so rich in prehistoric remains as Cambridgeshire. Every age is represented, and it seems right, therefore, that a separate chapter should be devoted to the earlier part of our subject, while Roman and Saxon and later ages may claim a chapter of their own.

PRE-ROMAN.

In early ages when the rivers were running at a higher level than now and the gravel beds were being laid down here and there, and mud was being deposited elsewhere, and the surface soils were creeping down the hillsides and being arrested in terraces on their flanks, man lived here and trimmed pieces of flint to meet the requirements of his rude life. These flint implements are the only record we have of his sojourn here and we speak of the implements, or the age, or the men as *Palaeolithic*, which means *ancient stone*.

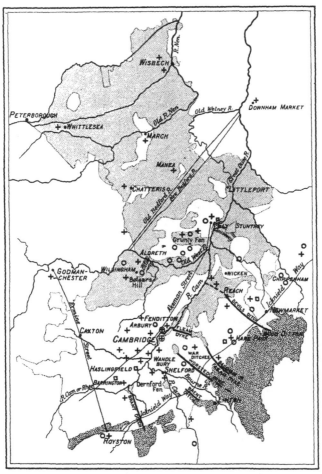

Map of Cambridgeshire showing the Dykes and the places
where antiquarian remains have been found

○ British.　　+ Roman.　　□ Teutonic.

▨ Woodland.　　▨ Fenland.

Some people think they have discovered implements of a ruder type, and, seeing in them the records of the *oldest stone* age, call them *Palaeotaliths* or *Eoliths* which means that they belong to the *dawn* of the Stone Age. The evidence for this is however very unsatisfactory.

In other parts of the world, where the Palaeolithic folk lived in caves, much more has been learned about them, for the remains of their food, bone implements of various kinds, and pictures drawn by them have been found, together with the bones of the same animals that lived with them here. The flint implements left by Palaeolithic man are found in great numbers in parts of East Anglia; there was evidently a manufactory of them in the adjoining county near Brandon, where flint seems to have been dug and dressed in every age. We know these implements to be the work of man because they show evidence of design in the way in which they are dressed all over by chipping, so that all sorts of rough stones are trimmed into uniform shapes. There are many different forms, but they all occur together here so that they do not indicate any difference of age. Elsewhere a rough succession has been made out from the fuller evidence obtained from caves, etc. Fig. 1 represents a fine specimen found just over the border at Mildenhall and another (Fig. 2) was found by one of the writers in the gravel on the top of the hill at Hare Park near Six Mile Bottom, 120 feet above sea level.

The stone implements of the Neolithic folk lie here and there in the soil and in the peat all over the district. They are not nearly so numerous as those of Palaeolithic

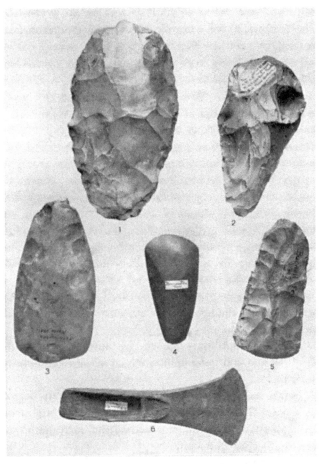

Prehistoric Implements found in Cambridgeshire

times and are made of flint or felstone or greenstone.
The felstone, as for example Fig. 4, and greenstone, and
mottled, or northern flint, as Fig. 3, are generally thicker
and of the bulging type; those made of the black Cambridge flint are thinner, more straight-sided, and chisel-
like (see Fig. 5). We cannot say whether this points
to unrecorded invasions from the north, or to the use of
stones from the drift, or to barter, or some other circum-
stance. Where stone implements of Neolithic Age were
made and all the chips and failures are lying about, we
see that in trying to knock out a Neolithic implement, the
maker very commonly trimmed it first into an oval or leaf-
shaped form, not unlike a Palaeolithic implement, and
worked it down afterwards into the straight or gently
bulging form, characteristic of the Neolithic Age. Then
he ground it to a good cutting edge and generally
polished all the surface, but as we never find one of
Palaeolithic shape ground and polished, it is clear that the
form had been changed before the custom of grinding the
implements had come in.

However that may be, we are dealing with a time
very long ago and geographical changes were going on all
the while.

Not a trace of the lion or hyaena, the grizzly bear or
the bison is found with Neolithic man. With him in the
peat and alluvium of the fens we find the boar and brown
bear, the beaver and wolf.

As we have seen in the chapter on Natural History
the urus was one of the few mammals which lived
on from Palaeolithic to Neolithic times, but it had dis-

appeared from Britain before the Romans came. That it lived here with the men who used polished stone implements is proved in the most satisfactory way, for the skeleton of one was found in Burwell Fen with a polished stone axe sticking in its skull (see annexed figure).

These Neolithic people practised inhumation, that is, they buried the bodies of their dead, placing them in the

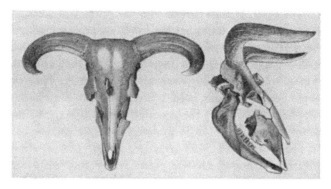

Urus (*Bos primigenius*) found in the peat of Burwell Fen

posture of sleep with the knees bent up and the head resting upon one hand.

Cambridgeshire had by this time assumed much the same geographical features as we see here to-day and we have no great changes to record when bronze was coming into use, although we have reason for believing that up and down movements were still going on.

The next people of whom we find traces here knew the use of metal, and, as copper proved too soft and the mixture of copper and tin known as bronze was the metal first commonly used, this is spoken of as the Bronze Age. Fig. 6 is a palstave or axe-head of bronze.

We must not say that bronze superseded stone for it was far too scarce and valuable and for some purposes stone was still used down to far later times.

Probably much of the woodland was cleared to provide fuel or the timber which was wanted for building huts on land, or even on piles driven into the bottom of lakes and swamps. It is curious that we have not evidence of these pile dwellings here, for the district is as suitable for them as that round Zurich, for instance, where they are found in great numbers. Moreover we know that the Neolithic people and those who used bronze and iron lived here as they did there, where their remains are found in the peat around the Lakes. It is probably because people have not been on the look out for these pile dwellings here, or have not known them when they have seen them.

Man of the Bronze Age burned his dead and buried their ashes in urns which are easily recognised by the conventional ornamentation on them. Surely we must suppose that these people had some idea of a future state to make them practise cremation at a time when sanitary reasons would not have suggested it. Over the grave they raised mounds of earth, or stone, such as those seen in Hare Park, in one of which the urn (Fig. 1, p. 146) was found.

We learn from history that different tribes held different parts of the country, and as some of them came from Gaul, some from Germany, some from Scandinavia,

they were not all alike when they were at home and so
they brought with them different arts of peace and war,
different appliances and different customs, and the study
of ancient earthworks and excavations here and there helps
to throw light on some of these obscure parts of our
history.

To the time of tribal expansion before the arrival of
the Romans we must refer those vast entrenchments which
crown the high ground along the principal routes into the
county—such as Ring Hill near Audley End just over
our boundary, Wandlebury on the Gog-Magogs, the War
Ditches by Cherryhinton, Arbury near Histon, and
Belsar's Hill near Willingham. These great circular
camps consist of one or more banks of earth with a ditch,
or ditches, outside, and, when strengthened by a stockade,
or palisade, must have been difficult places to storm.
Perhaps these were the *oppida* mentioned by Tacitus.
Some of them have been explored and their eventful
history partially revealed. As far as can be made out they
are all of British origin, but several show traces of having
been occupied by the Romans. For instance, in the War
Ditches above Cherryhinton there were fragments of
pottery of the rough British type at the bottom of the
fosse with bones of domestic animals. Above these a
number of skeletons lay as if thrown in and left un-
covered, the limbs and skull not always with the trunk,
and suggesting a massacre of young and old of both sexes.
The fosse showed signs of almost continuous occupation
from that time on, and the last few feet were full of
remains of fireplaces and cooking utensils and both rough

and highly ornamented pottery which some refer to late provincial Roman, some to late Celtic.

Tradition says that the circular camp known as Belsar's Hill was occupied by the Normans before they made their dash across the fens by Aldreth to storm the Isle of Ely.

But if we had to name the ancient earthworks which are most distinctive of Cambridgeshire we would point to the great dykes which cross the open ground between the high woodlands and the fens. They consist of a single bank with a deep fosse on the south side and sometimes a small secondary one on the north side also. They appear dimly at every stage of our history and we have occasion to refer to them in many connections.

As we enter the county near Royston we find one locally called the Brant (i.e. steep) Ditch running from the high clay-capped hills near Heydon to the once impassable tarn and swamp, known as Fowlmere.

Travelling north along the margin of the woodlands we come to another near Pampisford, which starts from the anciently wooded hills on the south-east and rests on the confluence of the Bourn and Granta on the north-west.

About three miles farther still to the north-east we find another immense earthwork running by Worsted Lodge along the open downs from above Linton to the Worts Causeway, well marked where the ground is high and dry and ceasing where it becomes swampy and impassable. This has been levelled on the top, and being a well marked track has been used as a road for ages and,

indeed, is commonly known as the Roman Road. Perhaps
the suggested explanation of the name Worsted, namely
Wool-street, may point to a time when pack-horses carried
over it the wool for which Stourbridge Fair was so long
celebrated.   Another dyke, also long used as a roadway,

Fleam Dyke, looking West

runs from the clay-capped hills of Balsham to Quy Water,
and a short one from the other end of Quy Water to the
river at Fen Ditton (Fen Ditch-town) completes this line
of defence.   These are both known as the Fleam Dyke
but it is more convenient to use the name Balsham Dyke
for the south-easterly one.

Then between Wood Ditton (Wood-Ditch-town) and Reach there is the tremendous bank and fosse which so impressed the old inhabitants as something beyond the work of man that they called it the "Devil's Dyke."

Perhaps we ought to include, as belonging to the same system of tribal defences, the Black Ditches just over our boundary south of Icklingham in Suffolk and the other "Devil's Ditch" running north from Brandon.

These dykes show what was the only open ground along which an enemy might advance and cattle might be lifted, which would be quite impossible through the tangled woods on the east, or the treacherous fens on the west. They thus mark out and defend what was the first nucleus of the county. Their uniformity of character leads one to think that they belonged to one tribe and the occurrence of the principal fosse on the south side would lead us to place the tribe on the north in the traditional home of the Iceni.

It is not clear when people began to make "dugouts," as the canoes formed by digging out the inside of a tree were called. There seems to be reason for thinking that the people of both the Neolithic and the Bronze Age lit a fire round the base of a tree or on the tree, and then hacked away at the charred part till they felled the tree and afterwards scooped it out into the required shape.

Besides the great earthworks which we refer to the Bronze Age, many scattered relics have been found which give us an insight into the conditions of life of those early ages.

The necklet or torque of twisted gold which was dug
up about a mile south of Witchford, suggests many
thoughts. How came this solitary memorial of a man of
rank and power to be left there in the wild fen? A
beautiful gold armlet was found in Grunty Fen. About
a mile west of Thetford a bronze sickle was found. A
mile and a quarter west of Stretham, a British urn was
turned up and with it the bones of the Urus. We have
seen that the Urus lived in Neolithic times (p. 137); does
this find prove that he was here in the Bronze Age
too? A mile south of Haddenham, by Hoghill Drove,
some gold ring money was found and about three-quarters
of a mile further east a great hoard of bronze spears and
palstaves and swords lay all together in the peat, in such a
manner as to suggest that a canoe with a cargo of bronze
"scrap" had been upset there. All these occurred in an
area not more than four miles each way, while objects of
the same age have been picked up here and there all over
the country.

There are no stone circles or alignments, no menhirs
or dolmens, in this part of England. It may be perhaps
because there is no rock that could furnish the stone, or it
may be because the builders of these monuments, which
can be traced from the Mediterranean up the West coast
of Europe, never got round to East Anglia.

Just before the coming of the Romans, a people on
the Continent who were evidently in touch with a high
civilisation as regards the ordinary appliances of domestic
life, brought or sent over a type of manufactured articles
to which the name Late Celtic has been applied. The

pottery much resembles Roman ware, and pieces of it have been found in this district associated with Roman remains (see p. 146).

ROMAN, SAXON, DANE.

We have now brought the story down to what must strictly be called historic times, for several Greek and Latin writers tell us something of the conditions of ancient Britain, and the historians of its conquest by the Romans describe what they found here. We must, however, read with judgment, for there was a great tendency to exaggerate the numbers and prowess of the enemy in order to magnify the achievements of their conquerors. They tell us that there were many different tribes and the Romans availed themselves of their jealousies to set them one against another and overcome them separately. We must bear these facts in mind when we endeavour to explain the position of the Romans in any district. In Cambridgeshire for instance, we find hardly any fortified stations. They do not seem to have won this part of the country by slow advance with exploratory or summer camps thrown up everywhere in front of them. They may have had to storm and occupy some of the *oppida* mentioned above but, when once they had broken the resistance of the natives in one or two pitched battles or by seizing all their strongholds, the British accepted the rule and adopted the mode of life of their conquerors. So that although for various purposes so much of the ground has been turned over and plenty of Roman remains found everywhere, they all tell of comfortable houses and farms scattered up and down our riversides and undulating downs in a manner that shows as much security for life and property as we have to-day. There is one stronghold at Chesterford just over the county boundary, but there is nothing else of the kind for many a long mile except some small earthworks at Grantchester (not the conspicuous moat near the church). There does not seem to be any Roman wall yet found in the county except a bit of walling for the base of houses here and there. The dykes we have described were there before the Romans came and it has been suggested that these were the enclosures which so hampered the retreating Iceni that Ostorius was able to cut them off, the terrible nature of the defeat being indicated by the exaggerated and improbable statement that 80,000 of the Iceni were slaughtered. Hardly anything has been found in the dykes. When the cutting for the Newmarket railway was made through the Devil's Dyke, nothing whatever was seen along the old surface under the dyke, although Roman pottery occurred abundantly just beyond it and a Roman villa had been built apparently under its shelter between the dyke and the road to Reach. A fragment of an amphora and a coin of Vespasian were found *on* the bank but nothing *in* it.

There is some reason for believing that the Romans began to control the waters of the fenland both by making embankments against the sea near Wisbech and erecting raised banks, probably as roadways, across parts of the fen as, for instance, from Reach to Upware. They also made waterways, such as the Carr Dyke, from the south of the Fens by Cottenham and Peterborough to Lincoln. Of Roman towns we can only name one, that is Cambridge, but there is no proof that it was fortified. Small villages, clusters of houses, and detached villas were the distinguishing features of Roman life in Cambridgeshire. There are coins everywhere. Gold coins are rare; but the quantity of copper coins is enormous. Sometimes swords and spears are found, but rarely armour; also carts and horse trappings. Remains of houses occur here and there.

1. British. 2. Late Celtic. 3. Roman. 4. Saxon.

They consist of a stone or brick base with painted walls of plaster on wattle. There are also all sorts of kitchen utensils, such as copper and earthenware saucepans and stewing pots of various shapes and qualities. Drinking vessels occur (Fig. 3), often ornamented with figures of animals and leaves and flowers. Quite common is a beautiful ware never made in Britain, but largely imported from Gaul. This is a very useful relic as it is

the most distinctively Roman, and is so easily known
by its texture and colour, like red sealing-wax. It is
generally called Samian as it was supposed to have been
first made in Samos. It frequently has hunting scenes in
relief on it and often the name of the maker, followed
by M for *manu*, or F for *fecit*; or sometimes Off for
officina (workshop) is stamped upon it. And if the catalogue of what has been found is too long for our present
purpose, the list of the places where Roman remains have
been discovered in the county would extend entirely beyond our limits. All along the valley from Chesterford to
Cambridge and up the hills on either side traces of Roman
sojourn have been turned up : not merely a few potsherds
which may have been left by the wayfarer, but rubbish
pits which have been slowly filled with household ware
and kitchen refuse. Large amphoras were found near
Chesterford Park. To the east at Barrington and Haslingfield were their cemeteries; so they must have lived
somewhere near. From Trumpington to Cambridge
excavations for phosphate nodules or gravel showed that
the ground was full of Roman remains. On the top of
the hill above Shelford clunch pit, and by Cambridge
Railway Station, pottery, coins, etc., prove their sojourn.
So if we follow the river to the north by Barnwell,
Chesterton, and Fen Ditton, to Biggin Abbey and
Horningsea, where numerous bronze vessels of Roman
date were found; or out to the north-east by Reach or to
the north-west by Madingley and Girton, the record is
the same. On the Castle Hill a vessel full of silver double
denarii was found by the clay diggers, and near Willing-

ham an urn full of small copper coins was turned out by the plough. At Upware a quantity of highly ornamented ware was found on the margin of the fen and along the ancient raised way, now buried under peat ; from there to Reach Roman urns were found buried.

Near Reach a Roman villa with a well-preserved hypocaust, i.e. a chamber for warming the house by hot air, was found close under the shelter of the Devil's Ditch. Near Landwade another was discovered with a beautiful tesselated pavement which has fortunately been saved and is now in the Museum of Economic Geology in Cambridge.

In the Isle of Ely and sometimes in the surrounding fen, Roman remains are common.

In fact wherever any extensive excavations have been carried on, e.g., for gravel, or phosphate, or fruit-growing, or building, Roman remains have almost everywhere been found.

It is interesting also to find settlements of Romanised Britons with their small boundary ditches occurring far out into the fen wherever a bank of gravel offered a dry site as at Histon, Willingham, Cottenham, Somersham, etc.

Probably things went on much as during the Roman military occupation for a while after home difficulties caused the last legionaries to be withdrawn from Britain, but at once Teutonic invaders came pouring in, and East Anglia of course was one of the first districts reached by them. They had been coming over both before and during the Roman occupation. The necessity for a *comes littoris Saxonici,* an officer whose special duty it was to protect the eastern districts from incursions from over the sea, proves that. But after the Roman soldiers and Roman municipal government were no longer there to organise resistance, the Romanised inhabitants soon fell under the domination of the strong Teutonic invaders. Then followed the long ages of small incursions which, in the aggregate, amounted to a vast invasion. Jutes, Danes, and Norwegians, and every variety of Scandinavian, or mixed Baltic race, landed on our shores. Angles, Saxons, Frisians and innumerable German tribes furnished their quota. And among the convenient landing-places the sands and strands round the Wash offered a safe and easy shore for running their vessels aground and hauling them up beyond the reach of the waves. The look of the people round the Norfolk coast tells us that this was their origin. These seafaring folk when they invaded our shores, did not always come with their families, but conquered and settled down among the pre-existing people.

The age of nomads and hunters had long passed away and a settled population had partly cleared the forests and reclaimed the land, thus producing by degrees changes in climate and many conditions affecting life.

The word Saxon has been used to mark the people of England of a particular time, not strictly as defining their original race, for although they severely repressed the older Romanised inhabitants they did not extirpate them, that is, kill them off root and branch, nor even exterminate them, that is, drive them away altogether beyond the bounds of their jurisdiction. Moreover there was another group of invaders, the stout-hearted Scandinavians, the later of whom we speak collectively as Danes, who contested the possession of England with the Saxons, and all England fell under the sway of one or other of these Teutonic races.

Though we know that these Teutonic races practically occupied all this part of England, we do not find anything like the same abundant traces of them as we did of the Romans, and almost all the objects we do find are associated with their cemeteries or sepulchral mounds. In tumuli on Allington Hill, near Hare Park, where centuries before the people of the Bronze Age had buried their dead, the Scandinavian settlers raised mounds over the bodies of their warriors, whom they laid in the grave with their spears and knives and gold-inlaid jewelry. The writhing dragons and other ornaments on the brooches found here, indicate their race and age. The brooch shown on the opposite page lay upon the breast of a skeleton in a tumulus (B) on Allington Hill, Hare Park, near Six Mile Bottom. It is circular, of bronze overlaid with gold, and is ornamented with an elaborate pattern of intertwined serpents or dragons. It bears five jewels, one central and four in the outer circle ; these jewels consisting of plaques of garnet on gold, set in beads of white glass paste. Not far off, the cemetery near Wilbraham, like that near Haslingfield, gives us another type which we call Saxon. The bodies were sometimes burned and the ashes collected in highly decorated urns (see Fig. 4, p. 146), and sometimes buried whole with indications of their rank and sex. The abundance of amber necklaces suggests that we should seek their origin along the Baltic or North Sea. The cruciform fibula represented on the next page is of bronze. Its surface is richly embossed in a variety of patterns and thickly plated with gold. (The pin, which was of iron, is wanting.) Its length is 7½ inches, and the width at top 3 inches. It is extremely

Brooch, from Allington Hill (Tumulus B)

large in size and of unusual beauty. It was found in 1879 in the Old English cemetery at Haslingfield. Associated with it were circular brooches, various brass plates which had evidently been attached to leather or wood ; beads of quartz, one of which was octagonally faceted ; beads of amber, blue glass, paste, etc., whorls of Kimeridge shale ; spear heads ; knives ; umbos of shields,

etc., etc.   These people have left no trace of their houses
and do not appear to have occupied the towns and villages
of the Romanised Britons whom they ejected.   Here and

Fibula found by Prof. Hughes in the Saxon
Cemetery at Haslingfield

there a dropped weapon or instrument and rarely a lost coin
tells us of their presence.   Possibly they adopted so much

of the appliances, coins, etc., of the Romans that we cannot now assign to each its probable owner. There was a mixed cemetery of Roman and Saxon in "the Backs" on ground belonging to St John's College from which a remarkably fine collection of urns was obtained, one of which is represented in Fig. 4, p. 146. Another cemetery was found at Girton, another at Barrington, and another at Wilbraham. At Orwell a coin of Offa, king of Mercia, was found and at Barrington one of Cynwulf. But these Teutonic people did not do much towards the embankment or drainage of the fens.

When the Normans came, we had something much more like the Roman conquest. They brought architects and engineers ; they built castles and cathedrals and had to attend to the waterways and roads and bridges to convey stone for their erection. The feudal or ecclesiastical lords established a monopoly in grinding corn and constructed mills and laid on water from the upper reaches of the rivers, to turn them. All this meant a control of the floods and of course certain geographical changes which were in some cases of considerable importance.

20. Architecture—(a) Ecclesiastical.

The fenland, which gave shelter and food to the ancient tribe of fenmen, drew Etheldreda back from Northumbria and made her forsake husband and queendom to found a Christian settlement away from the world, on the peaceful island, in her own country. The earliest

inscribed monument in the county to which we can give
a certain date, is the shaft and base of a stone cross
which now stands in Ely Cathedral ; it was brought from
Haddenham, where it had been for years used as a horseing
stone. In the seventh century it was put up there by, or
in memory of, Ovinus (Owen), who is known to have
been Over-Alderman to Etheldreda. The clear-cut Latin
words may be thus translated, "Give, O God, to Ovin
Thy light and rest "—they show the spirit of those early
seekers after truth.

An old monk of Peterborough, writing of the fenland
in 1150, speaks with delight of the islands "Which God
had raised for the special purpose that they should be the
habitations of His servants."

An abbey was built at Ancarig, or Thorney, as early as
657. Camden, quoting William of Malmesbury, calls the
island "A very Paradise bearing trees—that for their
straight tallnesse strive to touch the stars." There also
were apple trees and vines which he says " Either creepe
upon the ground or mount on high upon poles to support
them." "A wonderful solitary place is there afforded to
monks for quiet life." And many great religious houses
sprang up—for there was time to build them, money to pay
for them, and easy water routes by which to fetch material
from the great Rutland and Northamptonshire quarries—
with the hard chalk for inside ornament and the dark
oak for beams and roofs. The Danish invasions swept
the greater part of these away, but upon their ruins new
monasteries and abbeys sprang up. Canute favoured Ely,
and later the Norman Conquest revived the religious life

St Benedict's Church: Cambridge (Saxon)

of the land and gave a fresh stimulus and a new style to the religious buildings.

The period up to 1066, before the Norman Conquest, may be called Saxon, or Pre-Norman Romanesque. Its chief characteristics are " long and short work,"—stones at the corners of the tower and walls, built in flat and upright alternately—well shown in St Bene't's Church, Cambridge, with rough axe-work instead of the chisel-work which came in during the later Norman period. The arches are round or angular. An interesting group of windows and round openings, completes the top stage of St Bene't's tower.

From 1066 to 1170—roughly speaking—may be given as the Norman period. This style still shows the rounded arches and windows, plain massive columns, square abacus (the stone from which the arch springs); zig-zag ornamentation round windows and fonts, etc., as in the Round Church, Cambridge, and also billet moulding, such as is found round the windows of Ely Cathedral.

The characteristics of the Early English style, which followed for a century after this, from 1170–1270, are pointed arches often with " dog-tooth " ornament. The pointed arch was most probably invented to cope with certain difficulties of vaulting roofs with round arches, the use of the pointed arch giving the builder much greater freedom. The mouldings now became deeper, the abacus rounded, columns clustered (several narrower pillars being joined together to form one large pier, instead of the single massive round column of the Norman time). An

Arch, St Benedict's Church, Cambridge

example of Early English dog-tooth ornamentation may be seen in Jesus College Chapel, Cambridge, and in many of our churches.

Then, by gradual transition, came in the much more elaborate tracery of the Decorated style. Beautifully curved lines were seen in the windows, which became far more complicated in design than ever before. The abacus was now sometimes octagonal. The ornamentation was more elaborate. In the decorated parts of Ely Cathedral the ball flower ornament can be seen. The Black Death in 1348 put an end for a time to building, and with it to the Decorated style.

The Perpendicular style, peculiar to England, came after this and lasted until the Reformation. The lines now ran straight up to the top of the windows instead of forming a tracery of graceful curves, as in the Decorated work. The square effect of the straight transomes running across the windows, appears to be less ornamental than the earlier styles, but the windows of this period very often served only as a framework for the wonderful stained glass which was made at that time. Most of this has disappeared by now. In this country much of it was smashed during the Civil War by a Parliamentary agent called Dowsing, who at the same time destroyed all other ornaments and left the churches bare.

Of the Saxon work we have very little left. St Bene't's Church is our best example. A few fonts may go back to that date—one is seen at Guilden Morden. In Cambridgeshire early Norman work also is rare, though

some beautiful Norman buildings exist.   The struggle at
the time of the Conquest had exhausted the land, and the

Prior's Door, Ely Cathedral

reaction did not set in here until the late Norman and
Early English period had begun.

Ely Cathedral from the Ouse

NORMAN PERIOD.

The lower part of the tower and wings, as well as the nave of Ely Cathedral, are glorious examples of early Norman work. The Monk's and Prior's doorways, with their rich ornamentation, are of late Norman time. Other Norman buildings are Thorney Abbey, Stourbridge Chapel, belonging to the old leper Hospital, St Peter's church with its interesting font and the Round Church of the Holy Sepulchre in Cambridge. Unfortunately here the polygonal upper storey of the nave was "restored," that is destroyed, by the Camden Society in 1841, and imitation Norman windows[1], copied from an old one, took the place of the fifteenth century work. There are only three other round churches in England—at Northampton, Little Maplestead in Essex, and the Temple Church in London.

Wisbech Church has fine Norman columns and arches, and Tydd St Giles has good work of this time, and a tower of later date, separate from the church.

The interesting old priory church at Ickleton has some very early Norman work.

To this time belong St John's at Duxford; Hauxton Church—where there is a thirteenth century fresco of Thomas à Becket; part of St Peter's at Coton; the lower part of St Mary's tower at Swaffham Prior (St Cyriac's, standing in the same churchyard, has a tower square below, octagonal above—a later copy of St Mary's); the chancel screen at Rampton; and the one round tower of

[1] See pp. 162, 163, 225.

H. C.   11

N.W. View of St Sepulchre's Church, Cambridge

(Showing its condition at beginning of nineteenth century)

Cambridgeshire at Snailwell, with the double arch over the window, cut out of one stone.   This lovely little

St Sepulchre's Church: Interior

church stands on ground steeply scarped on the river side, as if it occupied the place of an old fort, commanding the fens.

Early English Period.

The Early English period shows that the country had recovered from its state of depression. Most of the churches in the county started with an Early English design, when there was a great increase of church building. Ecclesiastical affairs played a great part in

Ely Cathedral, Nave East

the life of the country, and much money was given for churches.

At this time the Galilee Porch at Ely was built, part of the cloisters at Jesus College, once St Radegund's Nunnery—the lovely double-arcaded chancel at Cherryhinton, Milton, Histon, Madingley, Stapleford, Barrington

—with a fine doorway—Little Abington, Foxton, Teversham, Bourn, Oakington, St Michael's, Longstanton, Littleport, and Haddenham.

None of these churches are Early English only; later additions and restorations have been made in every case. But much good work of the period remains.

DECORATED PERIOD.

To this time belongs the Lady Chapel at Ely, with its exquisite sculpture. The west doorway and the wonderful octagon lantern tower built after the old tower fell by that great genius, Alan de Walsingham. The structure of this octagon is unique. Alan of Walsingham is also said to have built Prior Crauden's Chapel and the Church of Little St Mary's, Cambridge, which has a beautiful east window. This was formerly the Church of St Peter, which was pulled down, the only part left being the Norman doorway. Trumpington is one of the few churches in the county built all in one style; it is pure decorated, and also has one of the oldest brasses in England.

Elsworth and Bottisham—the latter, which has been restored, has a stone screen—are also Decorated work. Grantchester has a good Decorated chancel. Over church shows fine window tracery, ball flower ornament and grotesque gargoyles; Decorated work is seen in Soham; Little Shelford; Fowlmere; Swaffham Bulbeck; Harlton (with a stone chancel screen); Haslingfield; and Willingham, which has a Decorated chapel, and a fine carved roof which probably once belonged to some other church.

Prior Crauden's Chapel, Ely

Balsham Church stands on the highest ground in the county, and has a fine Decorated chancel and tower. The old narrow doorway into the tower may go back to Saxon times, and remains of an old stairway are seen, possibly the very one where the last survivor of the Danish massacre held the foe at bay. In the chancel are two splendid brasses, misereres and a fine screen with a rood loft. The charmingly picturesque village of Horseheath has some fine monuments in its church and a beautiful brass which is believed by a local antiquary to commemorate a member of the Audley, not the Argentine family, as is commonly supposed.

Perpendicular Period.

There are many instances of Perpendicular work in our county. In Swavesey; in Sutton tower, standing high on the hill, one of the most remarkable in the county; in Soham tower with inlaying of black flints between delicate stonework; and Haslingfield tower, lovely in its position and proportions. Burwell Church, as a whole, is one of the grandest perpendicular churches in the county; Orwell, Chatteris, and Great Chishall, Outwell, Upwell, Whittlesea with its fine spire; March also with a spire and glorious carved roof, are all of this period; King's Chapel, finished in 1515, is one of the best examples of late Perpendicular work to be found anywhere. The style of its wonderful fan-vaulted roof, the culminating achievement of Gothic art in England, is peculiar to this country, and its organ screen has been

called the finest piece of carved woodwork on this side of
the Alps.   The glorious stained glass in all the windows,

King's College Chapel, from the West

except four, is English, and, from the College accounts, it
seems probable that it was saved during the Civil War by

a bribe administered to Dowsing, and by the fact that the Provost of that time was a Puritan. Many of the churches have fine oak roofs—notably March, Outwell and Upwell, Willingham, Leverington, Wilburton, Isleham, Elm. Some are carved with figures and angels, and some have plain oak beams intersecting one another, beautiful

King's College Chapel, from the South

in proportion and colour. Wall paintings also are not uncommon.

We have seen that Cambridgeshire was in all ways suited for the growth of abbeys and monasteries, which sprang up simultaneously in and round its borders. Ely and Thorney, Soham, and for a time Denny Abbey, belonged to the order of St Benedict. The Augustinian Canons lived at Barnwell with smaller "cells" at Anglesey and Spinney; Gilbertines at Cambridge, Fordham and Upwell; Hospitallers at Shingay and "Crutched Friars" at Linton, besides various houses in Cambridge. There were hospitals also—the one at Whittlesford still remains —for the care of the sick and infirm.

Many of the monastic houses were pulled down in the time of Henry the Eighth, and the materials were in many cases used to build up the colleges, which took the place of the monasteries in popular favour. The church of the abbey of Thorney was spared, on the condition that it should become parochial. Shingay, during the reign of King John, when England was put under an interdict by the Pope, was the only consecrated ground in all this part of the country where people could be buried; now it has completely gone.

RENAISSANCE.

During the sixteenth century English architecture began to be influenced by the Renaissance, as the revival of art in Italy was called. We find classical details gradually creeping into use in the Gothic designs during the Tudor, Elizabethan and Jacobean periods, but the great change was due to the genius of Inigo Jones, who first mastered the spirit of the Renaissance architecture and adapted it to English needs. This apparently entirely new idea of architecture was, as is so often the case with new ideas, only a return to a much older source of inspiration. For our Norman style had been developed, through the French Romanesque and Byzantine styles,

from the older Roman and Greek architecture.   The
Chapel of Peterhouse (1630) is an interesting example of
early Renaissance architecture in which the influence of
Gothic tradition is strongly marked.

The Chapel of Pembroke College (1664) is particularly
interesting as being one of the earliest works of Sir

The Great Court, Trinity College

Christopher Wren.   In this chapel we see the much
greater dignity and sense of proportion which were lacking
in the earlier attempts to blend new and imperfectly
understood ideas with the existing style.   Wren also built
the chapel of Emmanuel College.

21. Architecture—(*b*) Military.

The chapter on antiquities has shown that there are many great earthworks and lines of defence in Cambridgeshire which were thrown up in prehistoric time, and there are also many moated houses; but of military architecture belonging to historic times we have now hardly a wall left standing.

The early history of Cambridge centres round the Castle Hill. It is certain that the mound there is artificial, because it is made of chalk taken from the natural hill beneath it, and thrown up on the top of gravel and earth, quite out of its natural order. Some think that this mound was the base of a fort made by pre-Norman people, some that it was thrown up by the Normans. It rises on ground which was once occupied by the Romans and Romanised Britons. But whoever may have made the mound, we know that William the Conqueror built a castle here in 1068, on the site of twenty-seven houses which he pulled down to make room for it. These houses may have belonged to a town continuously occupied from Roman times, or rebuilt later under the protection of the old *Burh*, which we still see, crowning the height above Magdalene. This part of the town has been always known as " the Borough," and its burgesses were familiarly spoken of as " the Borough Boys," a name perpetuated in the sign of a public house in Northampton Street. The outer earthwork of the Norman Castle enclosed a large area. It can still be seen facing the

Cambridge Castle and Castle Hill from the Huntingdon Road

(*From an old print*)

river in the grounds of Magdalene College, and can be
traced across the corner of St Giles' Churchyard, enclosing

St Peter's Church, Cambridge

St Peter's Church and turning up under Pleasant Row.
The fosse belonging to this rampart was seen passing under

the Huntingdon Road where the Histon Road runs into it, and it turned down again towards the river, across ground now dug away for brick-making. In history we hear little about this castle; it was never taken, as far as we know, nor ever defended. It the early part of 1300 it was used as a prison, and afterwards as a quarry from which King's Hall, and later part of King's College, and

Sawston Hall

then Sawston Hall, were built. The keep, the most important inner part of the castle, was however still standing in 1553 and it was put into a state of defence by Cromwell's army in 1642. The Gate House remained until 1842, and now the County Courts and Gaol occupy its site.

At Ely there was probably a fort on Cherry Hill, for we read that the monks were required to find food in the abbey refectory for the garrison.

Picot, the famous Norman sheriff of the county, had a castle at Bourn, the moat of which still exists.

Wisbech had a Norman castle which enclosed a space of four acres, with a moat forty feet wide, over which there was a drawbridge facing west. An old seal is supposed to give a representation of it. It went through many vicissitudes, and was several times partially destroyed

Castle Ruins and Moat, Burwell

by inundations. It became the property of the Bishops of Ely. Bishop Morton rebuilt it in red brick as an episcopal residence, and Bishop Alcock died here in 1500. After the Civil War it was bought by John Thurloe, Secretary of State to Oliver Cromwell, who took the old building down and replaced it by an entirely new one designed in 1660 by Inigo Jones. It was still called the Castle, though it now became merely a private house,

After the Restoration it was given back to Ely, and was sold by the Bishop to a gentleman named Medway, who in 1816 pulled it down, and with part of the materials built a smaller house for himself.

At Burwell the moat and part of the wall of the old castle are still to be seen. It was constructed by Stephen to keep in check the outrages of Geoffrey de Mandeville, who was shot through the heart by an arrow whilst assaulting it.

St John's College Dining Hall

22. Architecture—(c) Domestic.

In pre-Reformation times much of the land of Cambridgeshire belonged to the Church, but almost every village has remains of a manor house which was often moated. These were the residences of ancient families whose monuments are seen in all the country churches. Many of the large mansions were pulled down, the contents and materials sold, and a smaller house was built on the site.

An Old Farm on the Cam near Haslingfield

The country people gathered together in villages from the earliest times. The old villages, some of them originally British, Roman, or Saxon, with their "common fields" and ancient church, contain many old cottages and some extremely beautiful half-timbered and moulded plaster work. The snug thatched houses are however disappearing as rapidly as the commercial zeal of advertising builders can replace them with up-to-date erections smeared with ornament. Many a little whitewashed cottage, too, now stands bereft of its thatch and covered with corrugated iron, painted red; like a person whose natural hair has been replaced by a tight-fitting skull-cap.

Queens' College, Cloister Court

But Whittlesford, Fowlmere, Horseheath, Triplow, Linton, Snailwell, and indeed nearly every village, still has beautiful old houses, some of them bearing an early date. Sutton, where some of the houses are built in steps on the steep slope of the old island, is like some little foreign town. Fine old farms with open chimneys and hams hanging in the chamber at the back of the fire, are still to be seen in the Isle of Ely.

Many grand old barns with upright timbers made out of whole trees, roughly trimmed and used head downwards, still remain in the villages of Horseheath, Wilburton, Snailwell and elsewhere. One can be seen near Shelley Row in Cambridge, where the French prisoners rested on their way to Norman Cross in Huntingdonshire.

In Cambridge itself beautiful old houses are constantly disappearing. The latest to go was the Falcon Inn, with part of an open gallery built round a yard, one of the many old hostels from which the street derived its name of Petite-curye or Little Cookery, now turned into Petty Cury. But many old houses still remain in Magdalene Street and Northampton Street, including the so-called "School of Pythagoras," which Mr Clark thinks was probably the country house of a Norman gentleman. It is built of stone and the principal rooms are on the first floor with the vaulted kitchens and cellars below, an arrangement characteristic of the twelfth century, from which time this house probably dates. Similar examples can be seen at Ely in the Deanery and Canon Kennett's house and part of the King's School.

The old "Three Tuns Inn," belonging to the sixteenth century, still stands opposite the Shire Hall. Here Pepys drank bumpers to the king and here Dick Turpin used to sleep when he came to Cambridge.

Other old inns are the "White Horse," with a hiding hole in the chimney of the sitting room, and

" The Cross Keys Inn " in Bridge Street, which has a fine carved oak mantelpiece and oak panelled rooms.

In the house at the south-east corner of the Market Place and Petty Cury, were some fireplaces of clunch beautifully carved with fruits and foliage, which have recently been removed to the Guildhall. The history of the house goes back to the fourteenth century, but it

Madingley Hall

was rebuilt by one of the Veysy family, whose initials I. V. with his trademark—the arms of the Grocers' Company—and the date 1538 appear on the mantelpieces.

To this period belongs Madingley Hall built by Sir John Hynde in 1543, and afterwards the seat of the Cotton family. The two names can be seen roughly scratched on the right hand stone of the mantelpiece in the hall. In this beautiful house King Edward the Seventh lived when he was a student at Cambridge.

Another most interesting country house, built a few years later, is Sawston Hall. It stands on the site of an older Gothic structure in which Queen Mary found shelter whilst a fugitive, after the death of Edward VI. When she had left the Hall and her pursuers arrived,

"Priest's Hole," Sawston Hall

the latter, in their rage at losing their prey, burnt the house. Mary saw the conflagration from a distance and is said to have exclaimed "Let the house burn, I will build Huddleston a better." But there is no proof that she ever did more than give the stone of Cambridge Castle for the purpose. The present owner is the direct descendant of the Huddlestons of Queen Mary's time.

The house is in the Tudor style with the long gallery

so characteristic of Elizabethan work. There are also a small private chapel and a priest's hiding place.

Chippenham Hall was built on the site of a Preceptory, and it was here that Charles I whiled away some hours of his captivity at Newmarket by coming over to play bowls. The present house is comparatively modern.

Childerley Hall, the seat of the Cutts family, is still standing. Charles I, when a prisoner, took supper and bed here on his way to Newmarket.

The original house at Impington Park belonged to the Pepys family, and was built by John Pepys in 1587. Some of the walls are between three and four feet thick, but though the ground plan is the same, not much of the original building remains.

Bourn Hall is an Elizabethan building which stands on the site of the old moated castle of Picot, the Norman sheriff of the county.

Trumpington Hall was formerly the home of the descendants of Sir Roger de Trumpington. The Pemberton family has lived here since 1675; Anstey Hall, also at Trumpington, was formerly the home of Christopher Anstey the writer.

Wimpole, now the seat of Viscount Clifden, is a large mansion of brick. The central part was built in 1632, and additions were made by the first Earl of Hardwicke in 1739. Stretching northwards for over two miles from the front of the house, is a magnificent avenue of elms in two rows.

Babraham Hall was the residence of Horatio Palavicini, who collected Peter Pence under Queen Mary and shared them with Elizabeth when she came to the throne. He is best known by his epitaph. The house in which he lived was one of the finest Gothic buildings in the county, but was in 1765 taken down.

Though most of the older country houses in Cambridgeshire have been replaced by newer buildings on the same site, we still have examples in our Colleges of what some of them were like. For the typical College plan and that of the mediaeval country houses were essentially the same ; both having the great hall as the nucleus, generally on one side of an open court, with the gate house on the opposite side, and in the case of the College, with the chapel on the third side. The hall, which occupied the whole height of the building, was entered from the court through a low passage called "the screens" with the hall opening on one side of it, the gallery, open to the hall, over it, and the kitchen and buttery opening off on the other side. At the further end of the hall was a dais, raised a few inches above the rest of the floor : it was lighted at one end, or both, by a bay window, and on this dais, in private houses, the family took their meals, as do the Dons in College now.

The large open fireplace was usually in the middle of the side wall of the hall, but in earlier days logs were burnt on the floor in the centre of the room. Beyond the dais, and approached through the hall, were all the private rooms of the house, which was thus divided into two entirely separate parts, between which the only access was through the hall. It was not until the seventeenth century that families began to prefer to have their meals in a small private room, and that the hall, instead of being the centre of family life, became a mere entrance to the house.

As that ideal handbook, Mr Clark's *Concise Guide to the Town and University*, is available for all, we shall only attempt here to give a few words on collegiate buildings in Cambridge.

Senate House and University Library

At Peterhouse, founded by Hugh of Balsham in 1284, the oldest of our Colleges, may be seen a few remains of the original thirteenth century building, such as the doors at each end of "the screens." The Combination Room and Master's Lodging over it were built in 1460. In the Hall and Combination Room are beautiful stained-glass windows designed by Burne-Jones and Madox Brown.

In Queens' College, founded about 166 years after

Peterhouse

Peterhouse, we have a good example of a College built on the plan of a manor house like Haddon Hall, with the long gallery and rooms of the President's Lodge arranged in the same position as those of the lord in the manor house. The view up the river, past the garden front of the President's Lodge, is one of the most beautiful in Cambridge. The wooden bridge was built in 1749. There was formerly another fine old wooden bridge, and the mill beyond it stands on the site of the King's Mill and Bishop's Mill, which were of so much importance to the town in early times.

When the Renaissance style came in, at the end of the sixteenth century, rich families and founders of Colleges were anxious to have buildings in the Italian style, and even pulled down much admirable work merely in order to be in the fashion. The Colleges seem to have vied one with another in building and re-building, or in modernising their existing structures by refacing them in the Italian style, with, in many cases, far from happy results. At first Classical ornamentation was merely applied to the Gothic method of building, which was eventually superseded.

Clare College, or, as it was first called, Clare Hall, one of the most beautiful buildings in Cambridge, shows a masterly blending of the two styles. Mr Clark tells us that the bridge was built by a master mason, Robert Grumbold, who probably designed, or at any rate suggested, the stone work.

The Fellows' Building in the second court of Christ's College is also an admirable piece of early Renaissance

Queens' College from the River

architecture, built in 1640–42, and from it an iron gateway leads into the secluded garden, with its fine lawns,
swimming bath, summer-house, and at the far end
Milton's mulberry tree.

The Lady Margaret Beaufort, mother of King Henry
the Seventh, founded St John's College as well as Christ's,
and the beautiful entrance gateway of St John's is

Christ's College, 1st Court

ornamented with her daisies and rose and portcullis. The
shield bears the arms of France and England, crowned
and supported by the Beaufort antelopes. The Old
Bridge, built in 1696, makes a charming picture, with
the view of the gabled red brick building of 1671 in the
background. The hall has fine oak panelling and the
original open roof. The beautiful gallery of the Master's
Lodge is now used as a Combination Room.

The Great Gate of Trinity College, built in 1518–35, commemorates King Edward the Third and his six sons. The blank shield stands for the little son, William of Hatfield, who died 1336 in infancy. The statue of King Henry the Eighth between the windows, with a live

The Combination Room, St John's College

sparrow usually to be seen on the sceptre and its nest in the crown, was added by Nevile in 1615. Sir Isaac Newton had an Observatory on the top of this tower. The west front of the gateway is later and commemorates the two visits of King James. The beautiful fountain in the Great Court was built by Nevile in 1602 and rebuilt in 1716. The charm and associations of this great square cannot, however, be crowded into one short sentence.

In 1676 Sir Christopher Wren built his magnificent Library at Trinity; in 1722 the Senate House and the Fellows' Buildings of King's were built by James Gibbs; and in 1758 the University Library by Stephen Wright.

Trinity College Library

Pembroke College Chapel and the Chapel and Cloister of Emmanuel College were designed by Sir Christopher Wren.

Emmanuel College, like Jesus and Sidney Sussex, stands on the site of a religious house; the ponds in the garden may be the very " pondes, stewes and waters " of the old Priory. The present Library was built by the founder, Sir Walter Mildmay, as a Chapel. He was a strong Puritan, as were other members of this College, one of whom, the Reverend John Harvard, emigrated to the English colony in Massachusetts, and dying there in 1638, left his library and half his estate to found a college "for the education of the English and Indian youth of

Emmanuel College

this country in knowledge and godlynes." Such was the origin of Harvard College, the earliest institution of learning in the United States.

We should hardly recognise Dr Caius' Gate of Honour, by which the student passed from his College to the Senate House to be honoured with a degree, if we could see it as it was originally. The stonework was painted white, the sundials, roses, and coats of arms were all gilt, pinnacles surrounded the hexagonal tower, and a weather-

Gate of Honour, Gonville and Caius College

cock, made in the form of a serpent and dove, rose from the apex.

In 1703 the Pepysian Library at Magdalene was probably finished, but no one knows its exact date or architect. Pepys' library arrived there in 1724, and the books still remain in his twelve bookcases of red oak.

Addenbrooke's Hospital is named after John Addenbrooke, M.D. The first building endowed by him was

The Fitzwilliam Museum

erected in 1766. In 1864–5 the Hospital was remodelled and greatly enlarged from designs by Sir M. Digby Wyatt under the directions of Sir George M. Humphry, but further additions are much needed.

The sum of £100,000 was bequeathed to the University in 1816 by Richard, Viscount Fitzwilliam, for building and maintaining a museum to contain his collections.   The architect who began the work, George Basevi, was killed in 1845 by a fall from scaffolding in Ely Cathedral.   He was succeeded by C. R. Cockerell, but the entrance hall, which was not finished until 1875, was largely modified by E. M. Barry.

In the first half of the nineteenth century came another architectural revival : this time in favour of

The Senate House

Gothic architecture.   Once again fine buildings were destroyed or mutilated to conform to the fashions of the time.   During this revival were built the Screen in front of King's, the New Court of Trinity by Wilkins, the New Chapel of St John's by Sir Gilbert Scott, and the gloomy Whewell's Court of Trinity by Salvin.

Since that time many good, bad, and indifferent

buildings have been erected in styles more various and more individual. Amongst modern buildings are the Law Library and Sedgwick Museum by Mr T. G. Jackson. The old thorn trees pointed out by Professor Maitland as marking one of the way-balks on the ground once known as "Swinescrofte," are still standing on the south side of the Museum. They are shown on Loggan's map

The Leys School

of 1688. During the building operations they were carefully protected. The New Medical Schools are by Mr E. S. Prior, the new buildings at King's and Caius College by Sir Aston Webb, Newnham College by Mr Basil Champneys, and Girton College and the new fronts of Caius and Pembroke are the work of the late Mr Alfred Waterhouse.

The Perse Grammar School was founded in 1615 by Stephen Perse, M.D., a Fellow of Gonville and Caius College. It was housed in Free School Lane until 1890, when new buildings, designed by Mr W. M. Fawcett, were erected in Hills Road. This famous school has done much for Cambridge and has produced many distinguished scholars. The Leys School, on a beautiful site between Trumpington Street and Coe Fen, was founded in 1875.

23. Communications: Past and Present.

In old times the chief means of communication were by the natural and artificial water-ways of which we have already given an account under the head of shipping.

There were, however, also certain routes leading through the country to the inland districts. Of these it is now difficult to find any direct evidence, and we should clearly distinguish between a route and a road. In order to realise this difference we must recall the conditions of inland traffic which must have prevailed in early times. The country to the north-east of Cambridgeshire was, we have reason to believe, rich in horses and cattle, large herds of which would be driven south and exchanged for other commodities, carried home on pack-horses. Now we know what this involved, from the customs which prevailed before the time of railways in, for instance, the remote parts of Wales. The cattle were driven up to the fairs and markets of England, and as it was impossible to carry fodder for say a thousand head of cattle, they had to find pasture as they went. The drovers, however, could not depend upon hiring suitable fields here and there as required on their long journey of two or three hundred miles. Therefore, by law or custom, there were rights of free pasture along certain routes. The belt of country travelled over was unenclosed and the cattle or horses fed and rested where necessary. As proprietors encroached on either side these routes became more and more contracted into roads, worn and broken by the trampling, now confined to a narrow track, or, if the way was mended, it was rough and hard, so that the oxen as well as the horses had to be shod to travel over them. The way in which cattle were driven within the last two centuries out of Wales, over Mynydd Epynt for instance, shows how cattle and horses must have been driven along the route, traditionally known as the Icknield Way, which traversed the only passable country between the fens and the high woodlands, from the country of the Iceni into the southern and midland districts. When the track was cut up in one place, they swerved right or left on to unbroken ground. When in later times the route was curtailed and limited by enclosures on either side, the track had to be repaired and that bit became a road. But the Great Icknield Way was not originally a road but a route, marked, may be, by many a path and track along its course.

Maitland, referring to much later times says, "I believe that most of the roads that entered Cambridge were but narrow tracks superimposed, as it were, upon the arable. But the Hills Road or Hadstock Way must have had broad strips of waste on either hand." It is interesting to note that this road makes straight for the Icknield Way.

An obviously ancient route from the south into Cambridgeshire was up the valley of the Dee and Stort by the great earthwork of Walbury, between Sawbridgeworth and Bishop's Stortford, and on past the strong British camp, now called Ring Hill, near Audley End, and so down the Cam valley, between the woodland on the east and the fenland on the west, across the country guarded by Wandlebury, the War Ditches, and Arbury, near Cambridge. The names by which these great strongholds are now known, show how entirely they have been lost to history or tradition.

There were also tracks or trodden-down paths into and through the woods and wastes—the straightest and shortest available route from one centre of population and trade to another, for we must not suppose that the relations between adjoining tribes were always unfriendly. When the Romans came they had at first to avail themselves of such facilities of travel and indications of directions as they found already in use. They could not construct a military or commercial road until the country had been subdued, and we must not expect to find in Cambridgeshire paved roads to guide us, such as we see in some countries where suitable stone was abundant, nor are there even roads "pitched" or paved with closely set cobbles. They did, where the ground was soft, lay down gravel, a mode of road-making in use ever since and brought to perfection and general use by Macadam, from whose name they are commonly spoken of as macadamised. It is therefore very difficult to obtain direct evidence of Roman roads in Cambridgeshire,

Worsted Street (Roman Road)

although we know that we are on some of the great lines from the Thames to the north, and fortified stations and villages and villas indicate the general direction of Roman advance.

From the south the Romans travelled along the old British route, above described, up the Lea and the Stort to Saffron Walden, which is full of their remains ; and then, leaving the British fort, Ring Hill, well to the left, they came down the Cam to Chesterford, which was a rectangular fortified town, more distinctively Roman than anything else in the district, and so on to the neighbourhood of Cambridge, where there is hardly a bit of raised ground along the rivers that does not show traces of Roman occupation.

Their Ermine Street, however, which became the Old North Road from London to York of coaching days, starting north along the same general line of country, seems to have followed another tributary of the Lea a little further west and, running by Royston and Godmanchester, to have crossed the Nene near Peterborough. There was probably a road in Roman times running from Colchester under the chalk hills by Bartlow to join the northern roads near Cambridge. Though the river may have been the principal means of transport north of Cambridge, it is probable from the distribution of Roman remains that there were more or less important roads running on either side of it to communicate with the Fen Road at Denver and join the road to Brancaster at Castle Acre.

In former days, before the country had been enclosed and levelled by cultivation, antiquaries saw here and there ridges which they took to represent portions of various Roman roads, and the coincidence of parish and county boundaries with some of these lines, confirmed

the idea that they were ancient roadways, unless we may
assume in some cases that the boundary was marked by
an earthen bank and ditch. In some cases, however, as
in that of the Fen Road from Denver to Peterborough
(Castor), the road was found in places to be made up
with three feet of gravel, and is generally believed to
have been constructed by the Romans.

The present lines of communication are largely based
on these ancient routes and roads, probably because the
same geographical features facilitated their construction
on the same lines, and partly, perhaps, from the distribution of population which arose out of the same causes.
At any rate the main road from the south and the Great
Eastern Railway follow the valleys of the Lea, the Stort,
and the Cam in the general direction of the Akeman
Street by Cambridge, Ely, and Downham Market to
Lynn. A road and railway run from St Ives by March
to Wisbech, and from March to Peterborough. A great
road now runs along the line of the Ermine Street from
Royston by Huntingdon, to cross the Nene into Northamptonshire near Winsford, and the straight modern
road from Cambridge to Huntingdon follows the supposed
line of the Via Devana. The roads of Cambridgeshire
are mostly straight, because the ground was unenclosed,
flat, and of similar character for very long distances.
When the road has been taken through an old village it
zigzags to avoid cutting across pre-existing enclosures, as
at Cottenham and Grantchester, but when a village has
developed along an ancient way the houses line the road
on either side as at Aldreth, where the village runs south

to the ancient bridge or other river-crossing, or at Haddenham where the village runs east along the ancient road from Earith to Ely. The Great Northern Railway follows a very ancient line of traffic along the Rhee valley, and from Royston runs at the base of the chalk hills to Ashwell, while the London and North Western Railway runs up the valley of the Bourn Brook and slips out of the county at Gamlingay. In the middle ages roads were not so much neglected as seems to be sometimes supposed and a suitable approach was generally constructed and kept up to the greater monastic and ecclesiastical establishments, and many of our modern roads are based upon these. The church path leading from the principal centre of population to the church is still seen in many country places, as in Barrington for instance. Since the time of railways the station has become the point towards which local roads all tend.

The Great Eastern Railway and its branches serve the greater part of the county. The main line, opened in 1845, enters the county at Chesterford, passes through Cambridge and Ely, then turns off to the east, leaving the county at Mildenhall. The line which runs on north to Lynn was not constructed till 1847. It enters Norfolk about nine miles north of Ely. The branch from Ely to March and Peterborough was finished in 1846, and in the following year branches were opened from March to Wisbech and from Cambridge to St Ives and Huntingdon. In 1848 the line from March to St Ives and the line from Newmarket to Six Mile Bottom were completed ; the intended extension of this line to

Chesterford was given up when the line from Six Mile Bottom to Cambridge was made in 1852. The embankment of the abandoned line may be seen on the east of the main line near Chesterford.

In 1852 the Great Eastern constructed the line from Shepreth to Shelford, but the Great Northern did not get running powers over this line into Cambridge till 1866. In the previous year the Great Eastern made a line from Shelford to Sudbury, entering Suffolk at Haverhill. The same Company completed their branch from March to Spalding in 1866, and from Cambridge to Mildenhall in 1885.

The Great Northern branch from Hitchin to Shepreth was made in 1851, Royston being the point where their trains entered the county. The London and North Western opened a line from Bedford to Cambridge in 1862, entering the county at Potton, and in 1866 the Midland Railway ran from Kettering to Cambridge and subsequently obtained running powers over the Great Eastern line from Huntingdon to Cambridge.

24. Administration and Divisions of the County.

Before the Norman Conquest, the country was divided up into townships, hundreds, shires, and kingdoms. Each township had its own organisation, some being under the control of a lord and some being free. The chief man was the reeve of the township—a position which would tend to grow in importance.

A number of townships made up a hundred, under a hundredman, who summoned the court, and who represented the hundred in the shire moot.

The shires consisted of several hundreds. Here the chief magistrates were the ealdorman, sheriff (shire-reeve; to distinguish him from the reeve of the township, a less important man), and the bishop. The church at this time played a great part in the affairs of the country. The organisation of the shires was largely modelled on the ecclesiastical system, which was started earlier. Church affairs came before the shire courts. The ealdorman was appointed by the national council (*witan*), and attended its meetings with the bishop. The sheriff secured the rights of the king in the shire, led the free-men in arms, and acted as judge in some cases. The ideas of justice at this time were primitive. In cases of murder the family of the murderer were generally made responsible and paid a fine to the family of the murdered man, who had a regular price put upon him in case of accidents of this kind, to be paid as atonement to his injured relations.

Cambridgeshire was first organised as a county under Edward the Elder, after he had reconquered the Danelagh in 912. He divided it up into hundreds and tithings, which correspond roughly to the present existing parishes. Every man had to belong to a hundred, whose members were collectively responsible for him.

In the ninth and tenth centuries there was a growing tendency towards feudalism. The great local lords were granted large compact territories, and their servants became responsible directly to them rather than to the king. The Norman kings checked this process by diminishing the power of the nobles, and by making the sheriff dependent on the crown. Roger Picot, sheriff of Cambridgeshire, was a tyrant in his time, owning large estates and exacting money for the crown. The county was poor after the Danish wars, and the land was confiscated by William for his Norman followers. These were given scattered estates, in order to prevent them becoming too strong in one place.

Ely, as has been shown in Section 2, was in a peculiar position, being directly under the abbot, not under the king. During the reigns of the first Norman kings, a continual struggle was taking place between the king and the nobles. The latter had their own manors and servants bound to serve on their land, and these serfs came under the jurisdiction of the courts of the manorial lords, not under those of the hundred, shire, or borough. Thus the regular courts of the county were hampered by this independent seignorial justice of the nobles. However, the country became centralised gradually, though Cambridgeshire suffered under Stephen from Geoffrey de Mandeville, one of the anarchist nobles. Justices were sent round by the central government, the system of trial by jury took the place of the old method of ordeal, and local administration was put on a firmer basis.

In the reign of Elizabeth, the justices of the peace acted as an important link between the Privy Council and the local parishes. They arranged the rate of wages according to the harvest, protected the poor, and acted as overseers in all departments in the county. Their duties have been taken over by the county councils in modern times.

Cambridgeshire is in an unusual position since it has two county councils—one for Ely, and one for the rest of the county. The former consists of a chairman, who is a justice of the peace, fourteen aldermen and forty-two councillors; the latter of sixteen aldermen, and forty-eight councillors. This body makes bylaws for the government of the county, inspects the administration, builds bridges and main roads, and appoints local officers.

The other administrative units of the county are the parish and the urban or rural districts. In Cambridgeshire, there are 168 parishes. These are managed by a parish council, consisting of five to fifteen members appointed for three years. The urban and rural districts are generally larger than the parishes, but they are no longer practical administrative courts.

The county is still divided up into hundreds as formerly. There are now eighteen in Cambridgeshire, four of these being in the Isle of Ely. The sheriff of the county attends the judges on circuit, and is the returning officer at parliamentary elections. He has to appoint an under-sheriff within a month of his own appointment. A remarkable feature about this county is that it shares a sheriff with Huntingdonshire. This is unusual, and may perhaps be a survival of an ancient union between the two districts in Mercian, or later times.

The Lord-Lieutenant, formerly the leader of the

forces in the county and an important official in the time of the Commonwealth, now represents royalty in the shire.

With regard to justice in the county, at the Petty Sessions not less than two justices or one magistrate make preliminary inquiry, and convict in minor cases. There are six divisions in Cambridgeshire for this purpose, and four in Ely. The Quarter Sessions are held four times a year. For the Assizes, the county is in the south-eastern circuit. They are held in Cambridge regularly, and also in Ely and Wisbech alternately.

The administration of poor relief, etc., in the county is minutely inspected by the Local Government Board, which appoints officers, audits accounts, and superintends the local authorities. The Poor Law unions generally correspond to the rural districts; Boards of Guardians are the local administrators of the Poor Law. The county returns three members to Parliament—for the Northern or Wisbech Division, the Western or Chesterton Division, and the Eastern or Newmarket Division. Two members represent the University and one the Borough. So it has six representatives altogether. The greater part of the shire is in the diocese of Ely, four parishes are in that of Norwich, and three of St Albans. The Bishop of Ely still exercises jurisdiction in the Isle of Ely, not the Archdeacon; this is a survival of the power granted to him as representing the abbot in Norman times.

25. The Roll of Honour of the County.

In considering the great roll of illustrious names connected with Cambridgeshire we can mention only a few of those who have left their mark upon the county, or have made the University famous.

More than 1200 years have passed since the death of St Etheldreda, Queen Abbess of Ely; and still her personality stands out clearly. We know her as a great lady, daughter, wife and sister of kings; as a great organiser who knew how to choose a faithful steward; as one who made a vow and kept it; who followed the light as she saw it, though it led her along difficult ways. Alan de Walsingham's corbels under the octagon at Ely tell her story, partly in fact, partly in allegory.

Abbot Simeon, then ninety years of age, began in 1080 to build Ely Cathedral and restore the Monastery. In the thirteenth century Hugh de Northwold completed the beautiful Presbytery, and when the great Norman tower of the Cathedral fell, Alan the Sacrist—the great architect known as Alan of Walsingham—faced the disaster and in six years replaced the ruin by his wonderful octagon lantern tower. To his genius Cambridgeshire owes much.

The first attempt to gather students together in a College was made by a Cambridgeshire man, Hugh of Balsham, Bishop of Ely, in 1280, when he tried to introduce scholars into a religious house, the ancient Hospital of St John. But the clerks and canons did not agree, and so, four years later, the Bishop placed them in houses by themselves, near the church then called St Peter's, now Little St Mary's, and from this church the first Cambridge College was named Peterhouse.

John Fisher, afterwards Bishop of Rochester, stands out as a great personality. He was friend and confessor to the Lady Margaret, mother of Henry the Seventh, and it was through his influence that she founded Christ's and St John's Colleges. He brought the great scholar Erasmus to Queens' College to teach Greek. Miles Coverdale studied at Cambridge and in 1535 made the first complete translation of the Bible into English.

Matthew Parker, Vicar of Landbeach, Master of Corpus, afterwards Archbishop of Canterbury, took a prominent part in the Reformation movement, but in Cambridge he is remembered chiefly for the magnificent collection of books and manuscripts which he bequeathed to Corpus Christi College. It was Parker's influence, backed by Queen Catharine Parr, which saved the Colleges from the fate of the monasteries and induced Henry the Eighth in 1546 to form Trinity College out of older foundations and to apply some of the confiscated Church funds to endow five Regius Professorships. Latimer, Ridley and Cranmer, who under Queen Mary suffered martyrdom for their protestant principles, were all Cambridge men.

Sir Robert Bruce Cotton, lived at Hatley St George and was buried at Conington in 1631. He founded the Cottonian library which is now in the British Museum, and saved many of the manuscripts dispersed from the monasteries.

In 1626, nearly a hundred years after Dr Caius took
his degree, when Milton was an undergraduate at Christ's,
when George Herbert was Public Orator, and Thomas
Fuller, the witty historian, was a student at Queens',

Jeremy Taylor

Jeremy Taylor entered at Caius College. He was born
at Cambridge, the son of a respected tradesman whose
father, Cranmer's chaplain, was martyred under Queen
Mary. Ordained at twenty-one, he gained the favour of
Archbishop Laud by his eloquent preaching and through
him obtained a Fellowship at All Souls, Oxford. He

died in 1667, in his fifty-fifth year, "sliding towards his ocean of God and of infinity with a certain and silent motion." His writings were numerous, but he is specially remembered by his books of devotion, *The Golden Grove*,

Thomas Hobson

and *Holy Living* and *Holy Dying*, and still more by his holy life.

Thomas Tenison, who became Archbishop of Canterbury, was born at Cottenham in 1636. He was vicar of St Andrew's, and when the plague was raging in Cambridge his courage and skill were of great service.

A powerful personality which made itself felt in Cambridge, before the dark days of the Civil War, was that of the famous carrier and Puritan Mayor of Cambridge, Thomas Hobson, who has given the phrase "Hobson's Choice" to our language. He saw that money was to be made by supplying hackney horses to the hard-riding scholars, so he kept a stable of "forty good cattle" with boots, bridles, and whips ready for use. Every customer was obliged to take the horse whose turn it was to go out—"That or none" was his answer to every objector. He became a rich man and left some of his money in trust for charitable purposes in Cambridge and for the maintenance of Hobson's Conduit, though it was not built at his sole charge. He died in 1630 at the age of 86, when his daily occupation was interrupted. No inscription marks the spot where he lies in St Benedict's Church, but Milton wrote two epitaphs "On the University Carrier, who sickened in the time of his vacancy, being forbid to go to London, by reason of the Plague."

Oliver Cromwell is connected by many ties with Cambridge and the county. He was a fellow-commoner of Sidney Sussex College; lived for some time at Ely, on property which he inherited from his mother's family; represented Cambridge in the Long Parliament, and made Cambridge the centre of the Committee for the Associated Counties.

At Hildersham fine brasses commemorate the Paris family, and Matthew Paris, the famous historian, was born there about 1200. He wrote until 1273, at St Albans, where he succeeded Roger of Wendover, and he is the last and greatest of our monastic chroniclers. Not only did he write fully on European and national matters, but he also added stories which show the condition of the people at the time. From him we hear of King John's misfortune in crossing the estuary of the Nene. He also wrote an account of each year's weather, and, being an artist, he illustrated his own works. Fully realising his own position he wrote, " The case of historical writers is hard, for if they tell the truth, they provoke men, and if they write what is false they offend God."

For more than three hundred years the Earls and Dukes of Bedford have owned large estates and have been benefactors to the fenland people. In the seventeenth century French refugees, seeking safety from persecution, settled in Thorney and founded a colony there under the protection of Francis, Earl of Bedford, who employed them in draining the fens.

In the year of the Restoration, 1660, Samuel Pepys, then aged twenty-seven, began his famous Diary, and for nine years, writing in cypher, he confided to it his inmost thoughts and private doings, together with the Court gossip of the time. He was educated at Cambridge and was connected with the county by family ties. In his later years Evelyn was his firm friend. He became Secretary to the Admiralty and President of the Royal Society. On his death he bequeathed his six manuscript volumes of diary and 3000 books to Magdalene College. In 1825 his cypher was made out and the "Diary" was published by Lord Braybrooke.

A practical philanthropist of whom Cambridgeshire

may justly be proud is Thomas Clarkson, the originator of the anti-slavery movement. He was born at Wisbech in 1760. At Cambridge he took the first place as Latin Essayist in 1785, when the subject proposed by Dr Peckhard, Vice-Chancellor at that time, was, "Is it right to make men slaves against their will?" Clarkson not only won the prize, but in collecting material for this essay gained such a knowledge of the atrocities of the slave-trade that he determined to devote his life to its abolition. He worked with the Society of Friends, and won the sympathy of Wilberforce. In 1807, the bill for the abolition of the African slave trade was carried, but Clarkson continued his efforts up to 1834 when slavery was done away with in all our colonies. Near Ware, an obelisk marks the spot where he knelt and vowed to give his life to this cause.

Francis Bacon (1561–1626) the founder of inductive philosophy, first directed the spirit of enquiry by seeking for science "not arrogantly within the little cells of human wit but humbly in the greater world."

Dr Barrow, Master of Trinity in 1675, was the teacher of Isaac Newton, the greatest of our natural philosophers and the discoverer of universal gravitation. Newton succeeded Barrow as Lucasian Professor of Mathematics; represented the University in Parliament; was knighted by Queen Anne and became President of the Royal Society.

For want of space we can mention the names of only a very few of those who have brought fame to the

University in the field of science during the nineteenth century ;—Stokes, who filled all the offices in Cambridge once held by Newton ; Clerk Maxwell, Kelvin, founders of the modern school of mathematical physics; Cayley, Sylvester, mathematicians; Adams, discoverer of Neptune; Sedgwick, one of the great founders of geology who spent his long life in the University and whose memory is preserved in the Museum raised by his influence and in his honour. Leonard Jenyns (Blomefield) passed the greater part of his life in Cambridgeshire and left observations and collections which show that he possessed that highest merit in a naturalist, absolute accuracy. Charles Darwin stands alone as the maker of a new era in Natural Science.

English prose is represented by Macaulay, the greatest master of historical description in our language, and by Thackeray the novelist.

E. H. Palmer, the remarkable linguist and Oriental scholar, who met his death in the Sinai Peninsula in 1882, was born in Cambridge and became Professor of

We must here, of necessity, leave out many notable contributors to the world's progress, statesmen, judges, divines, who are connected with our University. But amongst those who are "gathered to the Kings of thought" our poets are the peculiar glory of Cambridge.

Chaucer, whether he was a "Cambridge clerk" or not, knew Trumpington and its mill. Spenser's bright imagination wanders back in the *Faerie Queene* from count-

less adventures to our quiet streams, "the plenteous Ouse"
taking into his waters with many a river "the Guant."

> "Thence doth by Huntingdon and Cambridge flit,
> My Mother Cambridge, whom as with a Crowne,
> He doth adorne and is adorn'd of it,
> With many a gentle Muse and many a learned Wit."

Marlowe during his short life raised the Elizabethan
drama to the point where Shakespeare took it up: then came
John Fletcher, who wrote with Beaumont and Shakespeare; "rare Ben Jonson," playwright, player and poet;
Herrick, his literary son, who sang "of brooks, of blossoms,
birds and bowers"; George Herbert, Crashaw, Cowley,
and Waller, all Royalists; Milton alone in his grandeur;
Marvell, lyrist and satirist, who worked with Milton and
was loyal to him to the end. The pomp and wit of
Dryden came well from a Poet Laureate of Restoration
days. Gray, of the undying Elegy, was born in a cold
season, so that his genius never flowered freely, but
Wordsworth, Coleridge, Byron were poets who came
with the days of liberty and awakening, Edward Fitz-
Gerald caught and held the spirit of the East in his *Omar
Khayyám*, and Tennyson, with his heart full of nature,
went out to meet her new-found laws.

Milton at the age of twenty-three, when his seven
years at Cambridge were nearly over, wrote his lines of
self-dedication and was true to them all through his life.
In the twin poems written just after he left Christ's,
L'Allegro gives his joy in the country, but *Il Penseroso*
reflects the very soul of Cambridge. It is full of wistful
looking back to "the studious cloisters pale," to the

Statue of Sir Isaac Newton, Trinity College Chapel

glorious chapel, where he hears "the pealing organ blow, to the full-voiced quire below," and sees the heavenly vision. To the untimely death of Edward King, a Fellow of Christ's College and contemporary of Milton, we owe *Lycidas*, that elegy unmatched in English poetry. In Trinity College library lies the manuscript of *Lycidas*, of *Comus*, and of *Paradise Lost* in the dramatic form in which it was originally cast, all in Milton's own handwriting.

Fresh to Cambridge from the northern fells, Wordsworth finds the universal spirit in the new world of level fields, open sky, and moonlit groves. In his "nook obscure" at St John's, he sees from his pillow, "by light of moon or favouring stars,"

> "The antechapel where the statue stood
> Of Newton with his prism and silent face,
> The marble index of a mind for ever
> Voyaging through strange seas of Thought, alone."

In him, as in Milton, King's Chapel stirs new depths of feeling. Music casts before the eye "a veil of ecstasy" as "the soft chequerings of a sleepy light," "martyr, or king, or sainted eremite," fade into darkness and "every stone is kissed by sound."

> "They dreamt not of a perishable home
> Who thus could build."

A remarkable set of Cambridge students, amongst them Alfred Tennyson and Arthur Hallam, full of enthusiasm for the best literature of their own time as well as of the past, recognised the genius of Shelley, and, through them,

Adonais, his great tribute to the memory of Keats, was in 1829 first reprinted in England at our University.

The friendship of Tennyson and Hallam, formed in their brilliant college days, has given us that noble monument of devoted attachment "In Memoriam," which beyond anguish and darkness shows

> " That God which ever lives and loves,
>   One God, one law, one element,
>   And one far off divine event,
>   To which the whole creation moves."

King's College Chapel

26. THE CHIEF TOWNS AND VILLAGES OF CAMBRIDGESHIRE.

Abbreviations, &c.

The population at the last census is given in brackets after the name of the place.

S. = Saxon; N. = Norman; E.E. = Early English;
D. = Decorated; P. = Perpendicular.
ch. = church; m. = miles; Ry. = Railway; Stn. = Station.

The dedication is given in brackets after the first mention of the church.

Abington. There are three villages of this name supposed to mean the homestead of Abban, a man's name. Great Abington

Little Abington

(267) is on the south, and Little Abington (216) on the north of the river Granta, about 8 m. S.E. of Cambridge while Abington Pigotts or Abington-in-the-Clay is some 13 m. to the S.W. of Cambridge.

Babraham (308), a village about 7 m. S.E. of Cambridge and a hall of the same name with a ch. (St Peter's). The tower is possibly S. and there are some E.E. windows. There is a curious fresco in the north aisle.

Balsham (780), a village 10 m. S.W. of Cambridge near the top of the high eastern plateau of the county. It gives its name to one of the great dykes. The ch. (Holy Trinity) is mainly P. and contains some fourteenth century stalls, a handsome rood screen of the same date, and some fine fifteenth century brasses.

Barrington (499). The word means the village of the clan which had taken a bear as their badge and family name. It lies about 7 m. S.S.W. of Cambridge on a terrace of gravel and loam in which the bones of the elephant, hippopotamus, rhinoceros bison, urus, elk, lion, bear, and hyaena have been found.

The district was occupied by the British, whose settlement became afterwards the site of a cemetery from which a very rich collection of objects of the Romanised British and Saxons has been obtained. The ch. is E.E. with D. and P. additions.

Bartlow (254). A village 12½ m. S.E. of Cambridge. The two syllables mean the same thing (mound) in different languages and we learn from it that the great sepulchral mounds close to the village (but just on the county boundary) were to successive generations the conspicuous objects from which the place was named. It was at one time supposed, from finding dressed flints in the mounds, that they were pre-Roman, but in 1832–8 they were excavated and found to cover Roman interments. The ch. (St Mary) is chiefly interesting for its enormously thick walls, its round tower, its fresco of St Christopher and its position.

Bassingbourn (1234). A small town some 3 m. N.W. from Royston with a fourteenth century ch. (St Peter and St Paul), built of stone and flint, and containing monuments to the Nightingale and Turpin families, and also a library of sixteenth and seventeenth century theological books. The town seems to have been a centre of some importance at one time.

Bottisham (624)=Bodekesham, Bodec's enclosure, a large village 7 m. E. of Cambridge. It lies along the road to the Lode, and this and the river were the means of communication with the outside world before the races had given importance to the Newmarket road. The fine stone ch. dates from the thirteenth century with D. work in the chancel and a P. stone screen. Among the monuments of special interest are a canopied altar tomb in Purbeck marble to Judge Beckenham A.D. 1289 and various memorials of the family of Jenyns of Bottisham Hall.

Bourn (709), about 9 m. west of Cambridge, was early a place of importance, perhaps from its good water supply, for William the Conqueror granted the lands to one of his Norman followers who built a castle here of which nothing but the moat remains. There is a cruciform stone ch. (St Mary) in which the N., E.E., and P. styles are well represented.

Burwell (1974). This very ancient and interesting village probably takes its name from a *burh* or fort which may have stood where afterwards King Stephen built the castle of which we still see remains. Burwell, with its Norman buildings, and Newnham and its Saxon cemetery, may once have been separate villages, now they are joined by a picturesque raised way and avenue of trees. The ch. (St Mary) has Norman work in the tower; the interior work is in clunch, of which there are very extensive quarries close to the village. Early monuments and brasses carry it back to the end of the thirteenth or beginning of the fourteenth century.

Cambridge (38,379. This does not include members of the University, the last census being taken in Vacation time). The early history of Cambridge is a matter of inference rather than of direct record. The town grew here because this was the first high ground at the end of the principal waterway through the fenland from the sea. The earliest occupation of the site of which we have any proof is that by the Romans and Romanised Britons, whose remains, consisting of potsherds, etc., are common on the

St Sepulchre's Church, Cambridge

Castle Hill and on the slopes at the "Backs"—also about St Sepulchre's, Barnwell, and the Ry. Stn.—all these sites being dry gravel terraces.

H. C.   15

The taking and retaking of the town by various races, Saxons, Danes and Normans, tell of the importance attached to its geographical position. The Mercians seem to have occupied the Castle Hill while the East Anglians held the other side of the river. Danes seized it in 870 and "sat down there one whole year." Peace and prosperity came to the town and county under Edward the Elder and his successors, before the final conquest of the country by the Danes. Then the Normans used the fort upon Castle Hill and recognising the importance of the position took in the whole hill and built on it a castle "strong for situation, stately for structure, large for extent, and pleasant for prospect."

The influence of the great religious houses of the Fenland was early felt in Cambridge, encouraging a love of peace, and promoting the study of literature and art in the days of tranquillity before the first coming of the Danes, and later under the Anglo-Saxon kings before and after the second Danish invasion. Bands of scholars settled in the town, which gradually became a centre of learning.

In 1207 King John gave a charter to the burgesses which granted the town to them in perpetuity, gave them a Merchants' Guild, and the right to elect their own chief officer, first called a Provost then Mayor. So the town grew and flourished. Many fraternities were attracted to it, and its teaching must have become well known to induce so many students from Paris, Oxford, and elsewhere to migrate to it. The scholars lodged in private houses in the town and constant differences arose between them and the burgesses, until Henry III granted a Charter to the University, making special rules for the protection of the students. But the first provision for secular scholars, in a house of their own, was made in 1284 by Hugh of Balsham when he founded Peterhouse. The College system grew up from this beginning.

The two annual fairs dating back to the thirteenth century—Midsummer Fair and Stourbridge Fair—are still continued, though the glory has departed from them. The tolls of the

former were granted by King John to the Prior and Canons of
Barnwell, and of the latter to the lepers of St Mary's Hospital at

St John's College, Entrance Gate

Stourbridge, whose old chapel still stands not far from the fair
ground. Stourbridge Fair was one of the most important in

15—2

Stourbridge Chapel

Europe. From it Bunyan took the idea of his "Vanity Fair." It drew together an immense concourse of people and during the three weeks when it was open in September the town was packed to overflowing. The booths were built in rows like streets, one being called Cheapside. Each ware had its proper place. A signpost in Barnwell still points to Garlic Row. There was the Booksellers' Row, the Cooks' Row, the Cheese Fair, Hop Fair, Wool Fair, and the Duddery where £100,000 worth of woollen goods were sold in a week's time, and many thousand pounds worth of orders were usually taken by wholesale tailors from London, Norwich, and elsewhere. Goods of every kind were sold: it was here that Sir Isaac Newton, as a freshman, bought his famous prism.

After the commotion of the wholesale business and horsefair was over, the country gentry, from all parts, came in to spend their money and to see the puppet shows and performances. Up to quite recent times, within the last century, the theatre was very popular with both town and University and celebrated actors were to be seen on its stage. The Fair was opened with great ceremony by the Vice-Chancellor and University officials and by the Mayor and Corporation and Members of Parliament. This was one of the occasions on which friction arose between Town and Gown. The University was not at all popular at the Fair. Happily such feelings of rivalry and mutual suspicion have now given way to the wider view of a common good which is gained by mutual understanding and co-operation.

Castle Camps (713). The de Veres had a castle here which may have been on the site of the moated farm by the ch. (All Saints') a P. building in flint and rubble. This and the adjoining parish of Shudy Camps probably derived their names from the ancient earthworks in them.

Caxton (451). A village on the Ermine Street 9 m. west of Cambridge. The Petty Sessions for a large surrounding district are held here. The ch. (St Andrew's) is E.E. and P. in flint with stone facings and there is some fine Jacobean work in the George Inn. Here in former times was one of the wayside gibbets on which the bodies of criminals were left hanging.

Chatteris (4711). A market town of some importance, 12 m. N.W. from Ely. The name is spelt in many ways but its origin is unknown. Many objects of interest have been found in the neighbourhood, among them bones of the fossil elephant and other extinct animals, an urn with a large number of Roman coins, and various celts and swords of great beauty and interest.

Cherryhinton (2596). A good example of a village built along the ancient road instead of along the new straight road from Cambridge to Fulbourn. The site, like that of Fulbourn, was obviously chosen because of the abundant water-supply springing from the base of the chalk (see p. 61).

The ch. (St Andrew's), built in clunch and stone, is E.E. and P. Remains of a fresco can be seen on the wall of the east end. The parish of St John's on the Hills Road has been recently formed out of parts of Cherryhinton and part of Trumpington.

Chesterton (9591), is now merely a suburb of Cambridge. The district is full of Roman remains but there are no Roman buildings or earthworks visible. Near the Vicarage there is an interesting fragment of monastic buildings once held by a fraternity who looked after the property near Cambridge which belonged to the Abbey of Vercellis in France. The ch. (St Andrew's), with a beautiful spire, is a fine battlemented building in flint, mostly D. and P. but the sepulchral monuments suggest an earlier date.

Chippenham (including the hamlet of Badlingham 500). A village 5 m. N.E. of Newmarket. The stone ch. (St Margaret's) has some interesting wall paintings and a fourteenth century rood screen. A preceptory belonging to the Templars, and afterwards to the Hospitallers, was sacked in Wat Tyler's rebellion and on the site the present hall was built.

Comberton (419), a parish on the Bourn Brook 5½ m. W.S.W. of Cambridge. The ch. (St Mary's) is an E.E., D. and P. stone building, containing an octagonal E.E. font and the door and stairs of the rood loft. The district is full of Roman remains and a curious old maze, once on the village green, can still be traced in the school yard.

Cottenham (2393). A village, like Coton, said to have taken its name from the cots of the early settlers, 6½ m. north of Cambridge, on a gravel terrace rising out of the fens. It is full of objects of antiquarian interest. The Carr dyke runs across the fens on the north side and the Aldreth Causeway traverses the N.W. corner. Ditches, potsherds and other relics of the Romans and Romanised British are numerous and a moat in the village indicates the site of one of several important manor houses. The ch. (All Saints') is built of stone and rubble in the D. and P. style. Cottenham used to be famous for its cheeses, the manufacture of which has of late years given way to the cultivation of fruit.

Doddington (1340). Formerly a place of much greater importance and once the richest living in England. Now seven rectories have been carved out of it, namely Benwick, Doddington, Wimblington, March Old Town, March St Peter, March St John, and March St Mary. It is 4 m. south of March. The ch. (St Mary's) is an E.E. building in stone with a spire and some good carved wood inside, and has memorials to the Peyton, Richards, and Harding families.

Downham (1801), called Downham-in-the-Isle or Little Downham to distinguish it from Downham Market in Norfolk, is a village 3 m. N.W. of Ely. The ch. (St Leonard's) is Transition Norman and E.E. Remains of the magnificent palace built by Dr Alcock, Bishop of Ely, at the end of the fifteenth century may be seen in a farm house ½ m. from the church. Slight rising ground allows us to give as a derivation Down, a hill, and ham, an enclosure or homestead.

Duxford (685). Two parishes St Peter and St John were united. 9 m. south of Cambridge. St Peter's ch. is a N. and E.E. building. The other ch., St John's, is very similar. At Whittlesford Bridge there are remains of a small monastic establishment chiefly D. style seen in the Inn and the barn at the back of it.

Elm (1798), a large village in the Isle of Ely 2 m. S.S.E. of Wisbech. There is a fine E.E. ch. (All Saints') of stone. It owes its importance to its being on two main roads and the Wisbech Canal.

Elsworth (558), a small village near the county boundary between Cambs and Hunts 6 m. south by west from St Ives. An interesting geological formation is named from this village the Elsworth Rock (see p. 54). The ch. (Holy Trinity) belonged once to Ramsey Abbey. It is a D. and P. building of stone with a pinnacled tower.

Eltisley (352), a village on the borders of Cambs and Hunts 5 m. east from St Neots with a pretty green. The E.E., D., and P. ch., which once belonged to Denny Abbey, is built of stone, is dedicated to St John Baptist and St Pandionia the daughter of a Scotch king, who to escape the importunities of her suitors fled to the nunnery of Eltisley where she died and was buried by a well named after her. Her body was afterwards removed into the church (1344). St Wendreth also is said to have been buried in this church.

Ely (7706). Ely is the ecclesiastical centre of the county and diocese, a County Court district and Petty Sessional division and has a long and eventful history. Comparison with Cambridge shows a contrast rather than similarity. Cambridge grew where it is because it is on the highest navigable part of the river and on or near many of the principal roads and routes. Ely on the

Ely Cathedral, Tower from South Side

contrary was isolated and inaccessible, and therefore afforded quiet for study and contemplation, and a sanctuary where the hunted could easily elude their pursuers. There is historical evidence to show that, after the destruction of St Etheldreda's Abbey by the Danes, a college for priests was established here and down to the present time the Cathedral precincts have been called the College, not the Close. The political division known as the Isle of Ely extends much farther than the marsh-encircled hill which was called an island. We have elsewhere (pp. 24–30), in our description of the Fens, spoken of the geographical changes which were brought about by natural and artificial operations, and the commanding view obtained from the top of the Cathedral when the evening mists are spread like water over the fenland, helps us to realise what its surroundings were—when the people of the Bronze Age paddled their canoes along the unbanked rivers and over the flooded fen—when the Romans and Romanised British held out against the advancing English—when the Saxon Princess St Etheldreda founded the Abbey whose church grew into the Cathedral of the Normans—and we can follow the history of the great minster through the troublous middle ages until it became the peaceful centre of the well-drained agricultural district of to-day.

Fen Ditton (680) i.e. Fen Ditch-town, from its standing at the N.W. end of the Fleam Dyke which is thrown up across the rising ground between Quy Water and the Cam. The ch. is built of rough Barnack stone in the E.E., D. and P. styles. About a mile to the north are the remains of Biggin Abbey, a residence of the Bishops of Ely, and near here Roman remains have been found.

Fordham (1326) 5 m. north of Newmarket. The ch. P. built of stone and flint dedicated to St Peter with a chapel to St Mary. There was a Gilbertine priory dedicated to St Peter and St Mary Magdalene founded here by Henry III as a cell to Sempringham Priory, Lincolnshire.

Fowlmere (477, wrongly spelt Foulmire). The name is from *vogels meer* and was written Fugelsmare in Domesday Book. It is 5 m. south of Cambridge. The mere is now drained. The ch. is an E.E., D. and P. cruciform building of flint with a central battlemented tower. To the north of the ch. there is a circular earthwork called the *Round Moats*, probably British. Here, as in some other Cambridge villages, gallows were formerly to be found.

Fulbourn (1771) 5 m. E.S.E. of Cambridge. The name possibly from *foul bourn*, because the stream was so commonly turbid. There were formerly two churches in the same churchyard, one dedicated to All Saints the other to St Vigor, Bishop of Bayeux: but in 1766 the tower of All Saints fell, entirely destroying the building and all the old oak was stolen. St Vigor's is an E.E., D. and P. stone building with some interesting tombs and a fine brass to William de Fulbourn, chaplain to Edward III. Fulbourn Manor, the seat of the Townleys, is an interesting old brick mansion. The County Asylum with 700 inmates stands on rising ground south of the village.

Gamlingay (1722), a village on the west border of the county, 13 m. S.W. of Cambridge, with an E.E., P. cruciform stone ch. (St Mary's) having a battlemented west tower with small spire. It once had a good market but when the town was burnt down in 1660 the market was transferred to Potton.

Girton (498). A village 2½ m. N.W. from Cambridge. Now well known from the Ladies College, a handsome red brick building on high ground skirting the Huntingdon Road. The ch. (St Andrew's) is built of rubble and stone in the P. style. It at one time belonged to Ramsey Abbey. There are interesting fourteenth century brasses to former rectors.

Grantchester (1172) 2 m. S.W. of Cambridge, an ancient site full of Roman and Saxon remains. The strong earthwork near the ch. may be mediaeval. The ch. (St Andrew's) is a D.

and P. building of chalk and rag with stone ashlar and a battlemented tower at the west end.   The Bourn brook runs into the
Cam just above the village.

Haddenham (1686) i.e. the home of Hoeda.   A village
7 m. S.W. from Ely.   Within the parish is the hamlet of Aldreth
of which so much is heard in connection with the struggle between
the Saxons under Hereward the Wake and the invading Normans.

Girton College

Haddenham was part of the patrimony of St Etheldreda and seems
always to have been ecclesiastical property.   The ch. was founded
A.D. 673 by Ovin, steward to St Etheldreda, whose cross formerly
stood at the entrance to Haddenham.   Its inscribed base is now
in Ely Cathedral.   The ch. (Holy Trinity) is a stone building with
an E.E. tower.   An old mansion and dovecote still stand on the
site of the ancient Hinton Manor House.

Haslingfield (520).   The village is 5½ m. S.W. by S. from Cambridge.   At the moated red-brick manor house Queen Elizabeth spent the night on her way to Cambridge in 1564.   The ch. (All Saints') is a beautiful example of D. work, built of clunch and stone.   Extensive phosphate diggings were formerly carried on near the village and revealed a Saxon cemetery in which many very rich ornaments and other interesting objects were found (see pp. 151, 152).   The hill above the village,

Newnham College

called from its many chalk quarries, White Hill, is also known as Chapel Hill, from a famous shrine which once stood on its summit —that of "Our Lady of the White Hill"—to which people came in such numbers at Easter-tide that "the whole town of Hasling-field was scarce able to receive the pilgrims."

Hauxton (213) 1 m. N.E. from Harston Stn. was probably an important river crossing from Roman times.   The ch. (St

Edmund) built of clunch and rubble, N., E.E., D. and P., contains an interesting thirteenth century fresco of St Thomas à Becket.

Hildersham (187) 1½ m. N.E. of Linton Stn. The ch. (Holy Trinity) D. of rubble with E.E. western tower and early octagonal font, contains two remarkable oaken figures of Sir Robert Busteler and his wife, and several interesting monuments and brasses to the Paris family, who lived here in the fourteenth and fifteenth centuries and to whom Matthew Paris the historian belonged. This is a locality for rare plants. Fine gravel sections may be seen on the neighbouring hills.

Histon (980), 3 m. N.W. from Cambridge, had at one time two churches, St Andrew's which still exists, and has fine examples of E.E., D. and P., and St Etheldreda which was pulled down in the seventeenth century. There is a very strong moat near the church and present manor house. The large jam factory of Messrs Chivers, which employs over 1000 people, is in this parish (see p. 101), and much fruit as well as market-garden produce is grown.

Ickleton (598) 1 m. north of Great Chesterford Stn. The name probably has no connection with the Icknield Way though this ancient route crossed the river here. The ch. (St Mary) is a large N. stone building with central tower, tall spire, and sacring bell which hangs outside the steeple. The arcades of the nave are supposed to be N. There are ancient frescoes, rood loft and screen. Some remains of a Benedictine Nunnery founded here in 1190 by Aubrey de Vere, first Earl of Oxford, still exist and stone coffins have been found on the site.

Isleham (1600) 6 m. north of Newmarket. The ch. of St Andrew is a large D. stone building with a wooden roof with an inscription recording that Crystofer Peyton made it in 1495. There are almshouses founded by Elizabeth wife of Sir Robert Peyton, 1518, and the remains of an alien priory under St Jacutus near Dol, Brittany. The monks removed to Linton in 1254.

Spurgeon the great Baptist preacher was baptised in the river Lark close by.

Kirtling (616) 5½ m. S.E. from Newmarket. The ch. (All Saints') is a N. building in ashlared flint with a western tower supported by very large buttresses. Kirtling Hall, built in the reign of Henry VI but in great part pulled down and the remainder called Kirtling Tower, is the seat of Lord North. Princess Elizabeth was kept prisoner here before she became Queen.

Leverington (1124) 2 m. N.W. of Wisbech: now divided into three parishes—Leverington, Gorefield, and Southea-cum-Murrow. The ch. (St Leonard and St John) is a fine E.E. and P. building in Barnack stone with a battlemented tower and spire. There are sedilia, old stained glass, and several interesting monuments in it, and close by the ch. the high sea bank, locally attributed to the Romans, is seen.

Here, at Parson's Drove, woad is grown and the old woad mill may still be seen working. Parson's Drove is a chapelry in the parish; Fitton End is 1 m. north of Leverington.

Linton (1530). The word may mean "flax (linum) enclosure." A village on the Granta 11 m. S.E. from Cambridge. The ch. (St Mary's) is a N. and P. building in flint and rubble with a battlemented western tower. An alien priory of Crutched Friars, attached to the convent of St Iago de Lisle in Brittany, was founded here in 1292. Barham Hall or Priory, now stands on the site. A Roman villa was found on the other side of the river and a cemetery was cut through, in making the Cambridge-Sudbury Railway.

Littleport (4181) 5 m. north of Ely, its ch. (St George) an E.E. stone building with a finely-proportioned battlemented western tower which has an opening through the base, for passengers along the footpath. There is a large shirt factory here employing about 300 hands and many improvements have been recently carried out both in the town and the surrounding agricultural district, where the fenland has been skilfully drained and improved by "claying." Some of the races for the Skating Championships take place here.

Lode (Bottisham Lode) (659). This is a village which grew at the landward end of the lode, with a station called Bottisham and Lode. In the parish are the remains of an Augustinian Priory founded by Richard de Clare in the reign of Henry I or by the king himself, and dedicated to the Blessed Virgin and St Nicholas,—probably a dependency of Barnwell Abbey. It is now a private house known as Anglesey Abbey.

Long Stanton, made up of two parishes All Saints' (343) and St Michael's (93), the latter, a beautiful E.E. building, is of rubble and flint, built about 1230, with a double bell-gable at the west end. There is an interesting church chest in oak of the twelfth century. The roofs of the nave and porch are thatched. There was a palace of the Bishops of Ely here at which Queen Elizabeth stayed in 1564, and it was here that the ill-judged performance of a play, which had been intended for King's Chapel, was presented before Elizabeth and the undergraduates were severely rebuked by the Queen. The way in which the village straggles along the road points to its having been built on an ancient north-westerly thoroughfare.

Madingley (183) 4 m. N.W. from Cambridge. The ch. (St Mary) is an E.E. and D. stone building containing the Norman font from St Etheldreda's at Histon (see p. 238). The hall is a beautiful Tudor mansion in red brick built in 1543 by Sir John Hynde, admirably restored by its present owner. Here King Edward VII lived when he was an undergraduate.

Manea (1361). A small town on an island of gravel in the middle of the Ely Fens, near the Old Bedford River, 8 m. S.E. from March. The ch. (St Nicholas) is a modern building on the site of an old ch. It is said that Charles I intended to build a house here, and a mound close by is known as "Charlemont.' Bones of the urus and other relics have been found in the neighbourhood.

March (7565). The word is said to mean a mark or boundary. It is an important town in the centre of the northern or Fenland division of the county, a County Court district, and Petty Sessional division, situated on the river Nene, by which goods are conveyed northward to the sea, southward to Cambridge, and south-west to Bedford. It is now divided into four parishes. A causeway bordered by fine old elms leads up to the ch. of St Wendreda (1346) which has a beautiful carved oak roof. The tower stands on arches with a passage through, as at Littleport. The other churches are St Mary's, St John's and St Peter's and the Chapel of Ease of St Mary Magdalene, West Fen, and these have had parishes recently assigned to them. Sepulchral urns and Roman coins have been found near the town. The market is held on Wednesday. There are extensive corn mills and windmills for grinding corn here, also engineering and agricultural machine works.

Melbourn (1515). A large village 3 m. N.E. from Royston. The ch. (All Saints') is a D. and P. building in flint, with memorials to the families of Hatton and Hitch. This had evidently been an important district not only in Roman times but also through the middle ages. Traces of moated and other residences occur along the valley and its alluvium is full of remains of ancient domestic life. In 1640 Melbourn was the scene of a successful resistance to the collection of ship money. It has always been a stronghold of the Free Churches. The old Congregational Chapel was founded in 1694.

Mepal (324) 7 m. west of Ely on the Old and New Bedford Rivers which are crossed here by bridges. The new cut, known as the Hundred-foot River and its connections, are navigable to

Lynn in one direction and to Bedford in the other (see pp. 44, 120). The ch. (St Mary's) is an E.E. building in flint and stone. The road, now taken over by the County Council, formerly a private way, was constructed by General Ireton during the Civil War, to convey troops from Chatteris to Ely.

Milton (471). Not from Mill and town, but from Middletown, i.e. halfway between Cambridge and Waterbeach. The ch. (All Saints') is a N., D. and P. building in rubble and stone, with a western tower. Milton's family is said originally to have derived its name from this village. The Rev. W. Cole the antiquary and friend of Horace Walpole, lived here. He bequeathed to the British Museum 100 volumes of MSS, which have never yet been printed, containing valuable antiquarian matter relating to Cambridgeshire.

Newmarket (10,688), head of a County Council district 13 m. N.E. from Cambridge. The old market was at Exning, the ancient capital of East Anglia (see p. 124) and was probably transferred in consequence of the outbreak of the plague in Exning in 1227. This was a great country for horses, it may be even from the time of the Iceni, and Exning and Newmarket were well situated for horsefairs. Perhaps out of the trials of horses in the fairs arose the races which in later times made Newmarket celebrated. Its downs are admirably adapted for training horses as well as racing. The marvellous earthwork known as the Devil's Dyke (see p. 142) runs across the Heath. The patronage of Charles II largely developed Newmarket as a racing centre and we may gather what the place was like in his time from the description given by Macaulay.

The main street divided the town into two parishes, All Saints in Cambridge, and St Mary's in Suffolk. Exning, which is full of Roman and Saxon remains, and part of the parish of Wood Ditton (the village at the S.E. end of the Devil's Dyke), are now included in Newmarket.

Newmarket

(From an old print)

Oakington or **Hockington** (465). The town of the tribe of Hoc, 7 m. N.W. from Cambridge. There is a large E.E. and D. ch. in stone with an ancient font, and some stone coffins of about A.D. 1350. The manor was given by Turketyl, Chancellor of King Edred, to Crowland Abbey in 946, was sacked by the Danes in 1009, and restored in 1018. After the dissolution of the monasteries it was bought by Queens' College in 1557.

Orwell (562), i.e. "the well beside the brink" (Skeat). The well, until recently, gushed out just below the ch. in the midst of a prehistoric entrenchment. There is a very large clunch pit above the churchyard. The ch. of stone and flint, E.E. with a fine P. chancel, has a roof rich with contemporary armorial shields, bearing the arms of many of the chief county families. Until 1870 the Orwell maypole stood in a clump of fir trees on the hill west of the road, the old Akeman Street.

Over (860), i.e. a bank or shore. On the Ouse 9 m. N.W. from Cambridge, a favourite spot for fishermen. The ch. (St Mary's) is built of Barnack stone (E.E., D., P.) with a western E.E. tower and an octagonal spire 156 feet high. The fourteenth century Sanctus bell, the rood screen, the gargoyles of extraordinary beasts and birds, and other features are all interesting. In 1004 A.D. the benefice was given by Ednoth Bishop of Dorchester to Ramsey Abbey.

Pampisford (301), 8 m. south from Cambridge. The south door of the ch. (St John Baptist) may be S. and there are also a N. font and a rood screen of later date. The ancient earthwork known as the Brent Dyke runs through the park of Pampisford Hall.

Prickwillow (1143), 4 m. N.E. from Ely Stn., lies partly in Suffolk and partly in Cambs. The name comes from the willows out of which they made "pricks" or skewers. It is on the river Lark, in the fen, and the ch. (A.D. 1868), the vicarage, and the schools are all built on piles. The hamlet Burnt Fen is 4½ m. distant.

Quy-cum-Stow (324) i.e. Cow-eye, or Cow island. The two livings of Quy and Stow were joined A.D. 1273 or perhaps earlier. Quy is a village 5 m. E.N.E. of Cambridge on the stream known as Quy Water, which rises from springs at the base of the chalk and flows through what was formerly a swamp. The Balsham Dyke runs up to it on the S.E. and the Fleam Dyke on the N.W., the marsh being thought a sufficiently impassable defence across the intervening ground (see p. 141). The ch. is a D. and P. dark grey stone structure standing on a commanding ite at some distance from the village.

Reach (Reche, or Ruin Reach) 6 m. N.W. from Newmarket. Here the huge earthwork, known as the Devil's Dyke (p. 142), after extending 10 m. from Wood Ditton, comes to an end in the en. Until the middle of the eighteenth century, when it was cut back, it ran through the middle of the village to the Lode. The side of the village on the north of the dyke is in Burwell parish and that on the south is in Swaffham Prior parish.

A raised road, now exposed only during peat-cutting operations, once ran across the Fen from here to Upware: this and the Roman Villa, found here, are described in the chapter on Antiquities. It is said that there were once seven churches in the township and this tradition points to the importance of the place when merchandise of all kinds was brought by boat to its Hithe on the oldest of the lodes communicating with the Cam at Upware. One of the four famous fairs of Cambridgeshire was held here and, though bereft of its glory, is still kept up. The Mayor and Corporation of Cambridge proclaim it on the first Monday in May. A small modern ch. stands on the site of the ancient ruined chapel of St Etheldreda.

Sawston (1699). The town of the Saelsings, a thriving village, the seat of an important leather and parchment factory and also of extensive paper mills (see pp. 103–106). The ch. (St Mary) is a N., E.E., D. and P. building in flint and rubble, with monuments to the Huntingdon and Huddleston families. Sawston Hall, a moated manor house, still the residence of the Huddlestons, is said to be built largely out of the material of Cambridge Castle given to the then representative of the family by Queen Mary, whom he had protected and assisted to escape.

Great Shelford (1085), i.e. shallow ford, 4 m. south by east from Cambridge. The ch. (St Mary the Virgin) was rebuilt in 1387 by Thomas Patesle who was then Vicar. It is a P. battlemented structure in stone and clunch with a western octagonal tower upon a square base. About a mile north of the village copious springs, known as the Nine Wells, gush from near the base of the chalk, and supply the water which flows by Hobson's conduit into Cambridge.

Little Shelford (441) is almost continuous with Great Shelford. The ch. (All Saints') formerly belonged to the family of de Fréville. It is a N., D. and P. building of stone with monuments to the de Frévilles and others. There is an octagonal font, several twelfth century coffin slabs, and what is said to be a Runic slab built into the wall.

Shingay or **Shengy** (42). A preceptory of the Knights Hospitallers of St John of Jerusalem was founded here in 1140. The last of the buildings were pulled down in 1697 and now only the rectangular moat remains to mark the site. "The Preceptory possessed (like all those belonging to the Knights of St John and of the Temple) the privilege of exemption from Interdict, even of the Pope himself, Shingay accordingly was the only place for miles round at which the dead could receive Christian burial during the years when this ghastly spiritual weapon was used by Innocent III against King John. Local tradition till quite recent days kept in memory the 'fairy-cart' (i.e. the feretorium or bier) which by night conveyed to Shingay the bodies denied sepulture elsewhere " (Conybeare).

Snailwell (186).   From Snail, the name of the river which rises here.   A village 2½ m. north from Newmarket, with fine trees and beautiful thatched houses and cottages.   There was once

Snailwell Church

a monastic establishment here of which the fish poⁱds and the ch. (St Peter's) alone remain.   The ch. has a round tower with a very early window (Early N. or S.).

Soham (4230), the ham by the depression: perhaps referring to Soham Mere, now drained.   It was once a place of great

importance, being connected with Ely (6 m. N.W.) by Stuntney
Causeway, and with the sea by Soham Lode which joins the river
Ouse at Ely.   It was even a Bishop's see before Ely.   The first
bishop was St Felix of Burgundy under Sigbert, King of East
Anglia.   He founded a monastery here which was destroyed by
the Danes in 870 A.D.   The ch. (St Andrew's) is a Transition N.,

Windmill, Swaffham Prior

E.E., D. and P. cruciform building full of interesting features,
having a western tower 100 feet high with fine flint work.

Stretham (1204) i.e. the hame on the street—a village on
the old Akeman Street where it meets the road to Haddenham.
It is 4 m. south by west from Ely.   The ch. (St James) is a fine P.

stone structure with a western tower and tall spire.   It contains a
good oak screen and several ancient monuments.   Near the ch.
is a wayside cross, with an octagonal panelled base and four niches
in the top.

Sutton (1420), south-town: 6 m. west from Ely.   A village
built on the highest ground of the Island of Ely, along the ancient

Remains of the two Churches at Swaffham Prior,
(*From a print of* 1806)

route from Ely to Huntingdon, having a very fine ch. (St Andrew's)
which with its tall octagonal tower, crowned by a stumpy spire
between 15 pinnacles, forms a conspicuous landmark.   It is mainly
of one style, Transition D., but there are parts, e.g. the south Porch,
said to contain remains of N. work.

Swaffham Bulbeck (706), the ham or hamlet of Swoet
which afterwards belonged to the Bulbeck family, while the

adjoining Swaffham belonged to the Prior. A Benedictine nunnery was founded here in the reign of King John, A.D. 1190, of which remains still exist. The ch. (St Mary's) is an E.E., D. and P. building.

The northern part of the village, known as Commercial End, goes back to the time before railways, when goods were brought by water along Swaffham Lode to the Hithe which, as at Reach, Burwell, and Bottisham, gave importance to the place. Near Hare Park, archaeological remains, ranging from Palaeolithic to Danish times, have been discovered.

Swaffham Prior (950), or Great Swaffham, 6 m. west of Newmarket. Remarkable for its two churches (N., E.E., D. and P.) still standing in the same churchyard. The ch. of St Cyriac and St Julitta was given to the Abbey of Ely by Brithnoth, in the time of Ethelred the Unready, and a new ch. was built on to the fifteenth century tower in the eighteenth century. St Mary's, partly destroyed by lightning in the eighteenth century and recently restored, has an interesting octagonal tower on a square N. base. The Huguenot family of Allix has been settled here since 1730.

Swavesey (899) 3 m. S.E. of St Ives. There was an alien Benedictine Priory, founded here in the time of William the Conqueror by Alan de Zouche, which was attached to the monastery of St Sergius at Angers. It was dedicated to St Andrew, and the very fine existing ch. was the Priory ch. It is an E.E., D. and P. stone structure with a western tower. Raised overgrown banks and ditches indicate the position of the ancient Priory.

Here sometimes the skating matches for the Championships take place.

Teversham (222), a village 3½ m. east from Cambridge. Ch. (All Saints') given in 991 A.D. by Brithnoth the East Anglian hero, killed at the battle of Maldon, to the monks of Ely. The existing ch. is an E.E. and P. stone building having octagonal piers with floriated capitals, sedilia, and oak screen and some early monuments.   Wren, uncle to the famous architect, and Bishop of Ely, was vicar here.   He was a disciple of Laud, and this ch. was especially roughly dealt with during the Civil War by Dowsing, the agent appointed by Parliament to purify the churches.

Skating for the Championship at Swavesey.
A. E. Tebbitt of Milton and J. Hiam of Strafford

Thorney (1781), Thorn-island; also called Ancarig; 7 m. N.E. of Peterborough. This is one of the most interesting districts in the Fenland: in A.D. 657–662 a monastery for anchorites was founded here by King Wulfhere of Mercia, son-in-law to Anna of East Anglia, in answer to the petition of Saxulf, Abbot of Peterborough.   After its destruction by the Danes in 870 it was

S.W. View of Thorney Abbey Church

restored A.D. 972 by Ethelwold, Bishop of Winchester, as a Benedictine monastery, and dedicated to St Mary and St Botolph. The buildings are seen by the ruins to have been of vast extent. The monastery was dissolved under Henry VIII: only the western end of the ch. was spared by becoming parochial.   At Thorney there was a great settlement of the French refugees in the seventeenth century, many of their monuments remain in the ch. and churchyard.   After the dissolution it came into lay

Thorney Abbey

hands, the spirited Earls and Dukes of Bedford carrying on and extending the reclamation begun by the monks.   In the nineteenth century they spent upon the estate a sum estimated at over £2,000,000.

Triplow or **Thriplow** (428), 8 m. south from Cambridge. The sepulchral mound of Trippa.   There are three conspicuous tumuli here, from which one might feel inclined to derive the name. The ch. (? St George or All Saints'), on the hill, is an E.E., D. and P. cruciform building. On Triplow Heath the Parliamentary army, under Fairfax and Cromwell, met and determined to make an armed demand for arrears of pay. They had secured the person of the King and from here, in June of 1647, they marched towards London.

Trumpington (1084). This village, mentioned by Chaucer in the first line of *The Reeves Tale*, is 2 m. south of Cambridge. The

Denny Abbey near Waterbeach

ch. (? St Peter's or St Mary and St Michael) is remarkable as being almost entirely in one style—Geometrical D.—about A.D. 1320. In the north chapel is one of the earliest brasses in England, that of Sir Roger de Trumpington who died A.D. 1289. In the church-yard is the tomb of Professor Fawcett, Postmaster General. Trumpington Hall has been the home of the Pemberton family for over 200 years.

Waterbeach (1277). The beach or shore of the water: a village by the Cam 5½ m. N.E. from Cambridge. The ch. (St John

Baptist) is an E.E. and P. building.   Here are the remains of the
Abbey of Denny built in 1160 by Robert, Chamberlain to the
Duke of Bretagne, as a cell of Ely, and dedicated to St James and
St Leonard.   From the Benedictines it was transferred to the
Templars, and in 1290 passed from them to the Minor Sisters of
the Franciscan order.   It was refounded in 1342 by Agnes,
Countess of Pembroke, for nuns of the order of St Clare.   Part of
the N. and D. church is built into the farm; the refectory is used

The Market Place, Whittlesea

as a barn, and many other parts of the ancient structures can be
traced.

Whittlesea (Urban District 3909.   Rural parish, including
Coates, Estrea and district, 3194).   11 m. west of March in the
northern part of the county.   An ancient town with a picturesque
covered market house in the square by St Mary's ch.   The two
parishes (St Mary's and St Andrew's) are united.   St Mary's ch.

is a N., D., P. stone building with a very fine perpendicular steeple, having crocketed pinnacles, a spire, and flying buttresses. The ch. of St Andrew is a D. and P. stone building, with a battlemented western tower.

St Mary's Church, Whittlesea

There are extensive brick pits in the Oxford Clay, which is not puddled as usual, but worked by what is called the semi-dry process. Quantities of bricks are made here and the industry gives employment to a large proportion of the population. The site of Whittlesea Mere, now drained, is about 6 m. off, in Hunts. A massive gold ring and many other antiquities have been discovered in the neighbourhood, and traces of a Roman road have been found near Elderwell.

Whittlesford (785), the Ford of Hwītel, a man's name: a village in the valley of the Granta 7 m. south of Cambridge. The ch. (St Mary and St Andrew) shows a long and eventful history, the earliest remnant being referred to S. times, while the twelfth, thirteenth, fourteenth and every later century are represented. On Whittlesford Bridge there was once a hermit's cell. One of the eleven religious houses that took the places of hospitals or almshouses in the neighbourhood of Cambridge, stood here: its chapel is now used as a barn, and in the inn, close to the station, there is a fine oak roof which is supposed to have been in the hospital.

Wicken (663), 4 m. south of Soham Stn., from *wīc* a village. The ch. (St Lawrence) is a stone building having a battlemented western tower with pinnacles. A mile north of the village are the remains of Spinney Abbey (A.D. 1216) a cell of Ely or a cell of the Augustinian Priory of Barnwell. Wicken Fen (see pp. 31, 75) lies to the west.

Great Wilbraham (456), the ham or home of Wilburh, a woman's name. 7 m. east from Cambridge. The ch. (St Nicholas) is an E.E. and P. structure of flint, having a battlemented western tower with pinnacles, a N. font and some ancient tombstones. On the N.E. is the S. cemetery from which a great number of characteristic remains have been procured and described by Lord Braybrooke (see p. 150). Little Wilbraham (348) is close to, and Six Mile Bottom, in the parish, takes its name from the 6-mile race course running along the low ground.

Willingham (1611), the home of the Wifelings. 10 m. N.W. from Cambridge, 1½ m. from Long Stanton Stn. The ch.

(St Mary and All Saints'), one of the finest in the county, is built of stone and rubble and has parts which have been referred to S. and N., but is chiefly D. and P. The roof is said to have been brought from Barnwell Priory. There are two canopied tombs, and a pentagonal D. font. The most remarkable feature is a small vestry on the north side of the chancel: the roof, with richly carved "beams," is all of stone. Not far off is Belsar's Hill a British camp supposed to have been occupied by the Normans under Belasis (see p. 140).

Wimpole (240), the pool of Wina, a man's name, 3 m. S.E. from Old North Road Stn. and 9 m. S.W. from Cambridge. Wimpole Hall built 1632 A.D., but much altered and added to since, contains a fine collection of paintings by old masters. There is a double avenue of elms leading up to it 2¾ m. long and 100 yards in width. The Akeman Street crosses it.

Wisbech (9808), Ouse-bech. A municipal borough, market town, and centre of County Council District. The river Ouse once ran by Wisbech, but the direction of principal outflow and the names have been changed and the stream running through Wisbech is now called the Nene, but the town still has water communication with Cambridge, Hertford, and London by the Wisbech Canal and the Ouse. It is an ancient and important seaport. Its chief trade is the import of timber and general merchandise. Fruit and flowers, also potatoes, asparagus, and mustard seed, are grown for sale in the surrounding country. Water is brought 21 m. from chalk springs near Marham in Norfolk. The ch. (St Peter and St Paul) is a N. and P. stone building with an interesting brass (A.D. 1401) and other monuments. There are few traces now remaining of the original Norman Castle. The town has a good museum, including a collection of Fen birds, and another of marine and freshwater fish from the Nene and the Wash. Thomas Clarkson, the great advocate for the abolition of the slave trade, was born here (p. 215).

Fig. 1. The Area of Cambridgeshire (551,466 acres) compared
with that of England and Wales

Fig. 2. The Population of Cambridgeshire (185,594) compared
with that of England and Wales (in 1901)

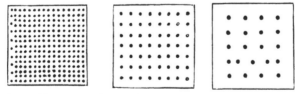

Lancashire, 2347   England & Wales, 558   Cambridgeshire, 212

Fig. 3. Comparative Density of Population to sq. mile (1901)

(*Note, each dot represents ten persons*)

17—2

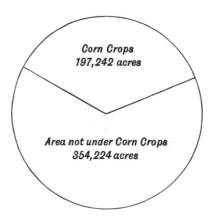

Fig. 4. Proportionate Area under Corn Crops in
Cambridgeshire

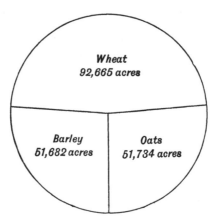

Fig. 5. Proportionate Area of chief Cereals
in Cambridgeshire (1908)

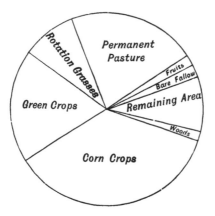

Fig. 6.   Proportion of Permanent Pasture to other
Areas in Cambridgeshire (1908)

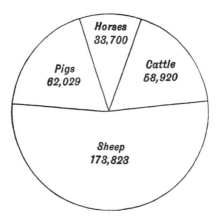

Fig. 7.   Proportionate numbers of Live Stock
in Cambridgeshire (1908)

INDEX

Abington 37, 39, 165, 222
Adams 216
Addenbrooke 194
Adventurers 30
Agriculture, School of 93
Akeman Street 22, 202, 244, 248, 258
Alban's, St 131, 208, 213
Alcock 231
Aldreth and Causeway 19, 20, 22, 44, 127, 129, 140, 202, 231, 236
Alien Priory 238, 239
Allectus 124
Allington Hill 150, 151
Allix 250
Ammonites biplex 57, 58
Ampthill Clay 52, 53
Ancarig 154, 251
Anglesey 169, 240
Anna 124, 251
Anstey 183
Arbury 139, 199
Argentine, 167
Ashwell 16, 17, 38
Assandun, Battle of 126, 127
Audley 167 and Audley End 199
Augustinian Canons 169, 240, 257

Babraham 37, 183
Bacon, Francis 215
Baitsbite 41
Balsham 16, 61, 167, 223
Balsham, Hugh of 185, 209, 226
Bardfield Oxlip 77

Barham Priory 239
Barnack 234, 239
Barnwell 66, 147, 169, 225
Barrington 18, 38, 64, 65, 114, 147, 164, 223
Barrow 215
Barry 195
Bartlow 37, 201, 223
Basevi 195
Basket making 22, 98
Bayeux 86, 235
Beaufort 189
Beaumont 218
Bedford 204
Bedford, Earls of 29, 30, 253
Bedford Level 30
Bedford Rivers 20, 44, 45, 241
Belsar's Hill 139, 140, 258
Benedict, St 169
Benedictines 238, 250, 253, 255
Benet's, St 155, 156, 157, 158
Benwick 47
Bernwood or Brentwood 96
Bentham 75
Bentley, Richard 108
Biggin Abbey 147, 234
Bishops Delf 12, 22
Bishop's Mill 187
Bishop's Stortford 35, 199
Bishop Moreton 46
Black Death 129, 158
Boat races 41, 42
Boadicea 123
Boards of Guardians 208

Book-binding 106
Bore, The 48, 49
Borings in the Backs 41
Borough Boys 172
Botolph 253
Bottisham 165, 224
Bottisham Lode 30, 122, 224
Bottisham Lode Cement Works 110, 111
Boulder Clay 24, 62
Bourn River 37, 39
Bourn Brook 37, 39, 231, 236
Bourn Village 39, 165, 176
Bourn Hall 183
Brancaster 201
Brandon 134
Brandon River 21
Braybrooke 214, 257
Brent Ditch 14
Bretons 86
Brithnoth 250
Bronze Age 138, 150
Brooch, Danish 150, 151
Bryggr, Bruggr 6
Brythons 83
Burh 172
Burnt Fen 21, 245
Burwell and Burwell Fen 32, 33, 34, 92, 114, 122, 128, 137, 167, 176, 177, 224
Byron 218
Byron's Pool 39

Caergraunt 5
Caius 192, 193, 211
Cam 5, 16, 17, 19, 35, 37, 40, 43, 45
Camboricum 3
Camboritum 3
Cambridge, origin of name 3, 4, 5
Camden 5, 16, 154
Camp of Refuge 19, 127
Camps 139
Canoes 142
Cantaloup Farm 39
Cantabrigia Depicta 117
Cantii 1

Canute 126, 154
Cardinal's Cap 109
Carr Dyke 145, 231
Carstone 57, 113
Castle Acre 201
Castle, Cambridge 128, 173–5
Castle Camps 229
Castle Hill 6, 147, 172, 173, 225, 226
Castor 202
Catwater 13, 47
Causeway, Stuntney to Ely, 248
Caxton 229
Cayley 216
Cells 169
Celtic short-horn 70
Cemeteries 147, 151, 152, 153, 257
Cenimagni 83
Chalk 52, 59–61, 112, 113, 114
Chalk, formation of, 59, 60
Chalk Marl 59, 61
Champneys 196
Chapel Hill 237
Charles I 131, 183, 240
Charles II 242
Chatteris 130, 167, 230
Chaucer 40, 216, 254
Cheesemaking 101
Cherry Hill, Ely 175
Cherryhinton 61, 164, 230
Chesterford 35, 145, 147, 201, 203
Chesterton 41, 230
Childerley 183
Chippenham 30, 34, 70, 75, 183, 230
Christ's College 187, 189
Clare College 187
Clare, de 240
Clare, nuns of St 255
Clark, J. W. Mr 185, 187
Clarkson 215, 258
Clifden, Viscount 183
Clunch 113
Cockerell 195
Cockle 21, 74

Coe Fen 30, 197
Coins 146, 147, 230
Colchester 201
Cole 242
Comberton 231
Comes Littoris Saxonici 148
Commercial End 122, 250
Conquest, The 127
Conybeare, Rev. E. 12, 247
Corbicula (Cyrena) fluminalis 66, 67
Coton 161
Cottenham 88, 101, 148, 231
Cotton family 181, 210
Count of the Saxon Shore 6
Cowley 218
Cox 130
Crab Hole 48
Cranmer, 130, 210
Crashaw 218
Cromwell 131, 213, 254
Cross Keys Inn 181
Crowland Abbey 244
Crutched Friars 170, 239
Cutter Inn 22
Cutts family 183
Cymry 83
Cynwulf 153

Danelagh 205
Darwin 216
Decoys 100
Denny Abbey 169, 232, 254, 255
Denver 44, 201, 202
Denver sluice 20, 44, 45
Dernford Fen 75
Devil's Dyke 113, 142, 242
Dialect 88, 89
Dick Turpin 180
Doddington 231
Domesday Book 4
Downham 231
Downham Market 202
Dowsing 158, 169, 251
Drift Map 62
Dryden 218
Dugdale 28

Duxford 35, 161, 232
Dykes 8, 17, 22, 77, 123, 124, 126, 140, 141, 142, 145, 234, 244, 245

Eager 49
Ealdorman 2, 205
Earith 20, 43, 44, 45, 203
Earl 2
Ecgfried 124
Edgar 9, 12
Edmund, St 126
Edmundsbury, St 116
Ednoth 244
Edred, King 244
Edward the Elder 205
Edward I 128
Edward III 190
Edward VI 130
Edward VII 240
Elderwell 257
Elizabeth, Queen 131, 206, 237, 239, 240
Elm (village) 169, 232
Elsworth 52, 54, 113, 165, 232
Eltisley 39, 232
Ely 9, 12, 19, 22, 101, 103, 114, 127, 128, 130, 154, 156, 158, 159, 160, 161, 164, 165, 166, 169, 180, 207, 232, 233, 234, 248, 249
Emmanuel 171, 191, 192
Eoliths 134
Erasmus 107, 210
Ermine Street 201, 202
Etheldreda 9, 19, 89, 124, 125, 154, 209, 234
Ethelwold 253
Evelyn 214
Exning 124, 242

Fairfax 254
Fairy Cart 246
Falcon Inn 180
Fawcett, Professor 254
Fawcett, W. M. 197
Felix, St 22, 248

Felixstowe 22
Fen Ditton 147, 234
Fen litter 75, 76
Fens, draining of 28–30, 131, 145
Fens 9, 18, 24, 257
Fen Road 201
Feretorium 247
Feudalism 205
Fibula 151, 152
Filey Brig 6
Finlander 86
Fisher 210
Fitton End 239
FitzGerald, Edward 218
Fitzwilliam 194
Fleam Dyke 141
Fleming 86
Fletcher 218
Flint knappers 114
Floods 29, 45, 47, 82
Fordham 170, 234
Fowlmere 165, 179, 235
Foxton 165
Franciscans 255
Free Churches 241
French Refugees 253
Freville, de 246
Frisians 84, 149
Fruticicola fruticum 66, 67
Fulbourn 75, 235
Fuller, Thomas 211

Gallows 129, 235
Gamlingay 56, 203, 235
Gibbet 230
Gibbs 191
Gilbertines 170, 234
Giles, St 174
Girton 147, 196, 235, 236
Girvii 2, 19, 83, 124
Gnawed bones 64
Godmanchester 201
Gog-Magogs 15
Gog-Magog House 16
Gold ring-money 143
Granta 4, 16, 35, 39
Grantchester 66, 145, 202, 235

Gravel 37, 57, 64, 82, 238
Gray 218
Great Bridge 5
Great Chishall 167
Great Ouse, Basin of 10, 35, 36
Grumbold 187
Grunty Fen armlet 143
Guilden Morden 158
Guyhirne 46

Haddenham 19, 143, 154, 165, 203, 236
Haddon Hall 187
Hadstock Way 199
Hallam 219, 220
Harding 231
Hardwicke 183
Hare Park 134, 138, 150, 250
Harlton 165
Harvard 192
Haslingfield 18, 38, 147, 167, 237
Hatfield, William of 190
Hauxton 30, 237
Haverhill 204
Henry III 128
Henry VIII 130, 170, 190
Henry of Huntingdon 16
Henslow, Professor 111
Herbert, George 211
Hereward the Wake 19, 236
Herrick 218
Hildersham 37, 238
Hinton Hall 236
Hinxton 35
Histon 101, 102 148, 164, 238
Hive, the 122
Hobson, Thomas 212, 213
Hockington 244
Horningsea 147
Horseheath 167, 179, 180
Hospitallers 170, 231, 246
Huddleston 182, 246
Huguenots 86, 250
Hundred 204
Hundred Foot River 44, 120, 121, 241

Hunstanton 18, 24
Hygre 49
Hynde, Sir John 181, 240
Hyth, Hithe 120

Iberi 83
Ice Age 19, 62, 63, 64
Iceni 7, 83, 123, 145, 242
Ice-scratched stone 63
Ickleton 35, 238
Icklingham 17
Icknield Way 14, 16, 22, 198, 199, 238
Impington 183
Ireton 242
Isleham 12, 169, 238
Isle of Ely 2, 12, 19, 234
Ives, St 54, 202

Jackson, T. G. 196
Jacutus, St 238
James, King 190
Jenyns 216, 224
Jesus College 158, 164
Jews 127
John's, St, College 40, 177, 189, 190, 227
John, King 128, 170, 226
John, regalia of King 128, 214
Jones, Inigo 170, 176
Jonson, Ben 218
Judges 207
Jura 53

Keats 220
Kelvin 216
Kemsworth Hill 15
Kettering 204
Kimeridge Clay 22, 52, 54, 55, 57, 113
King's College 175, 196
King's Chapel 130, 167, 168, 169, 221
King's Hall 175
King's Lynn 117
King's Mill 187
Kirtling 239

Kirtling Hall = Catlidge Hall 239
Knee-um 105

Lady Chapel, Ely 114, 130, 165
Lady Jane Grey 131
Landbeach 6
Landwade 148
Lark 44
Latimer 130
Laud 211, 251
Law Library 196
Leather 103
Leverington 169, 239
Leys School 196, 197
Lighters 120
Lincoln 24, 145
Lingay 39
Linnet 44
Linton 170, 179, 239
Lisle, de 239
Little Abington 165
Littlebury 35
Little Ouse 13, 21, 44
Littleport 21, 44, 78, 122, 165, 239
Little Shelford 165, 246
Little St Mary's 165
Local names of flowers 78
Lode 240
Lodes 9
Loggan 196
Long Brook 39
Longstanton 165, 240
Lower Greensand 22, 52, 56, 57
Lynn 117, 202
Lynn Law 30

Macadam 200
Macaulay 216
Madingley 147, 164, 181, 240
Magdalene 174, 194, 214
Maitland 196, 198
Maldon, Battle of 126, 250
Mandeville, Geoffrey de 128, 177, 206
Manea 20, 24, 240
March 12, 23, 24, 47, 120, 167, 169, 202, 241

Margaret, Lady 189, 210
Marlowe 218
Marvell 218
Mary, Queen 131, 182, 183, 211, 246
Mary's, St, Hospital 227
Maxwell 216
Maze 231
Medway 177
Melbourn 131, 241
Meldreth 38
Mepal 20, 45, 120, 241
Mercia 2, 84, 88, 124, 125, 207, 226, 251
Michael's, St 165
Middle Level 30, 44
Midsummer Fair 226
Mildenhall 203
Mildmay, Sir Walter 192
Milton (poet) 218, 219, 242
Milton (village) 72, 164, 242
Minor Sisters 255
Monasteries 129
Muscat 47

Nene 13, 44, 46, 74, 77
Neolithic 68, 83, 136, 138
Neots, St 61
Nettlefold Hill 15
Nevile 190
New Bedford River 44, 120, 241
New Leam 46
Newmarket 17, 61, 132, 242, 243
Newnham College 196, 237
Newton, Sir Isaac 190, 215, 217, 229
Nine Wells 246
Norman Cement Works 60, 61
Norman Cross 180
North Level 30, 44, 45, 47
Northwold, Hugh de 209
Norwich 208

Oakington 165, 244
Offa 153
Old Bedford River 44, 45

Old Croft River 44
Old Nene 47
Old North Road 201
Old Welney River 13
Old West River 20, 41, 44, 129
Oppida 139, 144
Orwell 167, 244
Ostorius 123
Ouse 12, 35, 36, 41, 43, 44, 77
Outwell 167, 169
Over 12, 44, 100, 244
Overcourt 43
Ovinus 154, 236
Oxford Clay 27, 52, 53, 54

Paigles 77, 89
Palaeoliths 68, 83, 132, 134
Palavicini 183
Palmer 216
Paludestrina 74
Pampisford 35, 244
Pandionia, St 232
Paper-making 106
Parchment 103
Paris, Matthew 213, 238
Parker 210
Parr, Catharine 210
Parson's Drove 239
Peat 32, 33, 34
Pemberton 183, 254
Pembroke College 171, 191, 196
Penda 124
Pensioners 130
Pepys 180, 183, 194, 214
Perse School 197
Peterborough 13, 46, 201
Peterhouse 171, 185, 186
Peter's, St, Churches 161, 165, 174
Petty Cury 180
Peyton 231
Phosphate 57, 59, 111, 112
Picot 4, 176, 183, 206
Pile Dwellings 138
Pirates 124
Pitt 109

Pleasant Row 174
Pottery, Roman, etc. 103, 146, 147
Potton 49, 204
Prickwillow 244
Priest's Hole 182
Printing 107, 108, 109
Prior 196
Prior Crauden's Chapel 165, 166

Queens' college 179, 188
Quy 30, 75, 245

Radegund's, St 164
Rampton 161
Ramsey 232, 235, 244
Reach 22, 113, 122, 142, 145, 148, 245
Records of a Fen Parish 120, 121
Reeds 100
Reeve's Tale 40, 254
Rennie 48
Rhee 16, 37, 38, 39
Richards 231
Ridley 130, 210
Ring Hill 139, 199
Ringmere, Battle of 126, 127
River terraces 68
Roman remains 132, 144, 145, 146, 147, 148, 149
Roman roads 17, 200, 201, 257
Roman snail 74
Roman villa 113, 145, 148
Roslyn Pit 22, 54, 55, 62
Round moats 235
Royston 11, 61, 131, 132, 202, 204

Sacring bell, 238
Saffron 96
Saffron Walden 35, 96
Salter's Lode sluice 47
Salvin 195
Samian ware 147
Sandy 56, 113
Sawbridgeworth 199

Sawston 35, 103, 131, 175, 182, 245
Saxulf 251
School of Pythagoras 180
Scott, Sir Gilbert 195
'Sea-Cole' 121
Seax 84
Sedge 75, 100
Sedge litter 75, 76
Sedgwick 216
Sedgwick Museum 196, 216
See, William 87
Sempringham 234
Senate House 185, 191, 192, 195
Septarian nodules 113
Sepulchre's, St 156, 161, 162, 163, 225
Sergius 250
Settlements 86
Sexburga 125
Sheep's Green 30
Shelford 37, 147, 246
Shelley 219
Shelley Row 180
Shells in gravel and clay 21, 64, 66, 67
Shells becoming locally extinct 74
Shepreth 204
Shingay 130, 170, 246
Ship-money 241
Shipping 117
Shire drain 47
Shire hall 180
Shire moot 205
Shire-reeve 1, 205
Shudy Camps 229
Siberch 107
Sigbert 248
Simeon, Abbot 209
Skating 28, 87, 240, 250, 251
Skegness 24
Smart, Turkey 87
Snailwell 163, 179, 180, 247
Soham 22, 167, 169, 247
Somersham 100, 148
South Ea 47
South Level 30, 44

Southrey 21
Spalding 204
Spenser 216
Spinney 170, 257
Spurgeon 239
Stapleford 37, 164
Steam pumps in fens 30
Stephen 128, 177
Stokes 216
Stonea 20
Stourbridge chapel 161, 227–229
Stourbridge fair 226–229
Stretham, and urn at 143, 248
Stuntney 22, 248
Sudbury 18, 204
Sutton 21, 167, 249
Swaffham Bulbeck 165, 249
Swaffham Lode 122, 240
Swaffham Prior 161, 249, 250
Swavesey 167, 250, 251
Swinescrofte 196
Sylvester 216

Tacitus 26, 139
Tawdry 89
Taylor, Jeremy 211
Telford 48
Templars 230, 255
Tenison 212
Tennyson 219, 220
Teversham 66, 165, 250
Thackeray 216
Thetford 41
Thetford bronze sickle 143
Thomas à Becket 161, 238
Thoresby, Ralph 49
Thorney 130, 154, 161, 170, 214, 251, 252
Thorp 84
Thousand willows 41
Three Tuns Inn 180
Thwaite 84
Timber, small area of 94
Tithing 205
Tonbert 124
Trinity College 171, 190, 191, 195

Triplow 131, 179, 253
Trumpington 40, 147, 165, 183, 254
Turketyl 244
Tydd Gote 47
Tydd St Giles 161
Tydd St Mary's 12
Tyler, Wat 129, 231

Unio litoralis 66, 67
University Library 191
University Press 107, 108, 109
Upware 22, 25, 41, 54, 148
Upwell 120, 167, 169, 170
Urns 143, 146, 148, 230, 241
Urus 70, 137, 143, 241

Vercellis, Abbey of 230
Vere, de 229, 238
Vermuyden 29
Veysy 181
Via Devana 14, 202
Vice Comes 2
Vigor 235
Villa, Roman 239, 245
Votiak 86

Walbury 199
Waller 218
Walsingham, Alan de 125, 165, 209
Wandlebury 16, 139, 199
War Ditches 139
Warship, first British 123, 124
Wash, The 24, 149
Waterbeach 6, 41, 254
Waterhouse, Alfred 196
Webb, Sir Aston 196
Webb, Jonas 91
Wedmore, Peace of 126
Welney River 44, 47
Wendover, Roger of 213
Wendred, St 23, 241
Whewell 195
White Hill 18, 39, 237
White Horse Inn 130, 180
Whittlesford 35, 170, 179, 232, 257

Whittlesea and Mere 23, 24, 27, 46, 47, 74, 167, 255, 256
Wicken 22, 31, 74, 257
Wilbraham 150, 257
Wilburton 169, 180
Wilburton bronze hoard 143
Wilfred, Archbishop 125
Wilkins 195
William of Malmesbury 154
William I 172, 206
William Rufus 87
Willingham 148, 169, 257
Wimpole 183, 258
Winsford 202

Wisbech 6, 20, 44, 46, 47, 48, 97, 117, 118, 119, 120, 161, 176, 202, 258
Witchford torque 143
Woad 97, 98
Wood Ditton 242
Wordsworth 219
Worsted Street 17
Wren 171, 191
Wright, Stephen 191
Wulfhere, King 251
Wyatt, Sir M. Digby 194

Zouche, Alan de 250

Lightning Source UK Ltd.
Milton Keynes UK
UKHW011146061022
410012UK00007B/241